T0271424

# Superconductivity Centennial

# Peking University–World Scientific Advanced Physics Series

ISSN: 2382-5960

**Series Editors:** Enge Wang *(Peking University, China)*
Jian-Bai Xia *(Chinese Academy of Sciences, China)*

Peking University-World Scientific Advanced Physics Series

Vol
6

# Superconductivity
# Centennial

Editor

## Rushan Han

*Peking University, China*

北京大学出版社
PEKING UNIVERSITY PRESS

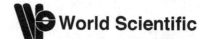
World Scientific

*Published by*

World Scientific Publishing Co. Pte. Ltd.
5 Toh Tuck Link, Singapore 596224
*USA office:* 27 Warren Street, Suite 401-402, Hackensack, NJ 07601
*UK office:* 57 Shelton Street, Covent Garden, London WC2H 9HE

**British Library Cataloguing-in-Publication Data**
A catalogue record for this book is available from the British Library.

B&R Book Program

**Peking University-World Scientific Advanced Physics Series — Vol. 6**
**SUPERCONDUCTIVITY CENTENNIAL**
Copyright © 2019 by Han Rushan

The Work was originally published by Peking University Press in 2012.
This edition is published by World Scientific Publishing Company Pte Ltd by arrangement with
Peking University Press, Beijing, China.

ISBN 978-981-3273-13-9

For any available supplementary material, please visit
https://www.worldscientific.com/worldscibooks/10.1142/11066#t=suppl

Desk Editor: Christopher Teo

Typeset by Stallion Press
Email: enquiries@stallionpress.com

Printed in Singapore

# Preface

It has been one hundred years since the discovery of superconductivity.

Superconductivity was discovered in 1911 by Kamerlingh Onnes when he was measuring the resistance of Hg at the temperature of 4 K. Below this transition temperature, the metal Hg has no resistivity, and it is called zero-resistivity effect. After years of research, in 1933, people realized another basic property of superconductors, the perfect diamagnetism, which is also called Meissner effect. Based on that, BCS theory was established in the 1950s, more than 40 years after the discovery of superconductivity, which shows the abstruseness of this theory and the difficulty of the work. There have been four Nobel Prizes given to 6 scientists because of their contributions to the superconductivity research. And the concept of pairing condensation has been expanded to 13 orders of magnitude in temperature, from $^3$He ($T_c \sim 10^{-3}$ K) to the nucleus ($T_c \sim 10^{10}$ K), without considering the quark condensate. BCS theory has affected the condensed matter physics and even the whole physics world over half a century. But the unusual features of the superconducting state and normal state of high $T_c$ superconductors challenge the authority of conventional BCS theory and even the Landau-Fermi liquid (FL) theory. In the last century, especially in the latter half, FL theory and BCS theory made important contributions to a deeper understanding of physics based on lots of experiments. In addition, the study of high $T_c$ superconductivity broke through the framework of FL theory and BCS theory on the shoulder of this "giant" again, and developed the condensed matter physics. At the same time of looking for materials with higher $T_c$s, the study of the mechanism of high $T_c$ superconductors attracts the attention of nearly the whole physics community. Although people have not reached a consensus until now, as a leading subject, the scope of the research field involved and the depth of the problems discussed in it are both rare in recent decades. It challenges the conventional condensed matter physics completely and promotes research greatly. It can be imagined and expected that the work in future will be difficult, and we will go on making efforts to explore. "Looking upon this unfinished scientific Tower of Babel" (P. W. Anderson).

In the dialogue between P. W. Anderson and R. Schrieffer in 1991 about the theory of cuprates (Physics Today, June 1991, P55), how to write the second volume of *solid state physics* was mentioned. They said like this: "Bednorz and Müller's 1986 discovery did mark the beginning of a remarkable period of development in condensed matter physics. Before that time, strongly correlated fermion systems were an interesting byway of the field, but most serious many-body theorists believed Fermi-liquid theory could cover the most interesting materials. We are now rewriting the condensed matter textbooks of the future by adding volume II, in which interactions must be included in zero order, on an equal footing with one-body kinetic effects ... but rather how to develop concepts and methods to handle such systems in general ... . Just as BCS was the dawning of a new type of physics now extending over 13 orders of magnitude in temperature, so we are perhaps witnessing the beginning of a major advance in our understanding of systems most of which are yet to be discovered." The two physics masters pointed out that high-temperature superconductivity physics would play an extremely important role in the development of physics.

A paper about Hg system high-temperature superconductor [Nature **363**, 56 (1993)] was selected as one of the most important ten articles in the twentieth century by Nature magazine, and included in Physics Century Anthology, which shows the full affirmation to the importance of superconductivity research and its important role in physics in the last century. The $T_c = 133$ K (it can reach 160 K under high pressure) of $HgBa_2Cu_3O_{8+x}$ reported in this article still keeps the highest record in the family of cuprates. There are Chinese scientists among its authors.

This book attempts to collect the important contributions in superconductivity research made by Chinese scientists, and combine them into a book as centennial of superconductivity. Because Chinese scientists participate in many important stages and problems, I collect these twenty papers in an appreciated and respectful mood, most of which have been published and others are articles engaged by special arrangement. This book is the companion book of another two books: *Advances in Theoretical and Experimental*

*Research of High-Temperature Cuprate Superconductivity*, and *High-Temperature Superconducting Physics*, the Second Edition. (Peking University Press, forthcoming in 2012.) *Superconductivity Centennial* is from the perspective of looking back at the history over twenty years, and *Advances in Theoretical and Experimental Research of High-Temperature Cuprate Superconductivity* is from the perspective of more comprehensive physical properties, while *High-Temperature Superconducting Physics* focuses on the special properties of copper oxides high-temperature superconductors. These three books complement one another. Of course they are more complement to the *Superconductivity Century Anthology* pressed internationally, among which some of the articles have been accepted, but we focus more on the contributions of Chinese authors in *Superconductivity Centennial*.

The works involved in the collected twenty articles are all on important problems of high-temperature superconductivity and are attached great importance by our peers. For examples, the paper about Y system high-temperature superconductors with $T_c = 90$ K by Wu Maokun and Zhu Jingwu's research group was the earliest report in the world of material with $T_c$ over liquid nitrogen temperature, the paper of Zhao Zhongxian and Chen Liquan's group reported this nearly at the same time; the research about Hg superconductivity in which Guo Jiandong participated still keeps the world record of superconducting transition temperature. The book also includes pairing symmetry research by Cui Zhangqi's group, NMR research by Zheng Guoqing's group, research of $d$-wave symmetry spectra by Ding Hong's group, measurement of the superconducting gap by Dai Pengcheng's group, the article about microscopic electrical heterogeneity by Pan Shuheng's group, the earliest research of ARPES by Shen Zhixun's group, the research of "Kinks" by Zhou Xingjiang's group. This book collects the important articles about iron-based superconductors too, such as the article by Chen Xianhui's group, the article which gives the highest superconducting transition temperature of iron-based superconductors by Zhao Zhongxian's group, the article about Zn doping in iron-based superconductors by Xu Zhuan's group, and the article about the spectra by Wang Nanlin's group. In order to fully introduce the physical properties of the condensed matter, this book collects the research about thermodynamic properties by Wen Haihu's group, the research about low temperature heat transport by Sun Xuefeng's group, and the research about Raman spectra by

Zhang Qingming's group. In addition, there are three important theoretical papers separately given by Ding Qinsheng's group, Su Gang and Guo Wei. Ding Qinsheng's group made the computing research to the magnetism and superconductivity of 122 iron-based materials. Su Gang discussed the off-diagonal long-range order in condensed quantum phase and reviewed the superconductivity, superfluids and Bose-Einstein Condensation, especially the discussion on super states. After pointing out the magnetic origin of the superfluid, Guo Wei suggested that the fluctuations in spin of O caused by hole doping in $CuO_2$ plane could lead to a resonating-valence-bond (RVB) state with quantum number $S = 1$, $S_z = 0$ and cause the magnetic ordering via the local exchange interaction which is called Kramers super-exchange. And the complete phase diagram was given from ferromagnetic insulator to superconductor.

In the 21st century, there is still a lot of work to be done about the mechanism of high-temperature superconductivity, especially the research of spin pairing mechanism. The key experiments like the isotope effect in conventional superconductors are needed. I think that we should search them from the special properties of high-temperature superconductors. Therefore I collect the special and maybe the universal properties in this system and compile them into a book, supply it to the readers for analysis and consideration, and I hope it could push the further research forward. I also wish the readers to give criticisms, modifications and supplements for this book.

*Han Rushan*
Peking University
July 2012

# Contents

# 1

# Superconductivity at 93 K in a New Mixed-Phase Y-Ba-Cu-O Compound System at Ambient Pressure*

M. K. Wu[1], J. R. Ashburn[1], C. J. Torng[1], P. H. Hor[2], R. L. Meng[2], L. Gao[2], Z. J. Huang[2], Y. Q. Wang[2] and C. W. Chu[2,3]

[1] *Department of Physics, University of Alabama, Huntsville, Alabama 35899*

[2] *Department of Physics and Space Vacuum Epitaxy Center, University of Houston, Houston, Texas 77004*

[3] *Division of Materials Research, National Science Foundation, Washington DC 20550*

A stable and reproducible superconductivity transition between 80 and 93 K has been unambiguously observed both resistively and magnetically in a new Y-Ba-Cu-O compound system at ambient pressure. An estimated upper critical field $H_{c2}(0)$ between 80 and 180 T was obtained.

The search for high-temperature superconductivity and novel superconducting mechanisms is one of the most challenging tasks of condensed-matter physicists and material scientists. To obtain a superconducting state reaching beyond the technological and psychological temperature barrier of 77 K, the liquid-nitrogen boiling point, will be one of the greatest triumphs of scientific endeavor of this kind. According to our stud studies [1], we would like to point out the possible attainment of a superconducting state with an onset temperature higher than 100 K, at ambient pressure, in compound systems

---

* Reprinted with permission from M. K. Wu, J. R. Ashburn, C. J. Torng *et al.*, Phys. Rev. Lett. **58**, 908 (1987). Copyright (1987) by the American Physical Society.

generically represented by $(L_{1-x}M_x)_a A_b D_y$. In this letter, detailed results are presented on a specific new chemical compound system with $L = $ Y, $M = $ Ba, $A = $ Cu, $D = $ O, $x = 0.4$, $a = 2$, $b = 1$, and $y \leqslant 4$ with a stable superconducting transition between 80 and 93 K. For the first time, a "zero-resistance" state ($\rho < 3 \times 10^{-8}$ Ω-cm, an upper limit only determined by the sensitivity of the apparatus) is achieved and maintained at ambient pressure in a simple liquid-nitrogen Dewar.

In spite of the great efforts of the past 75 years since the discovery of superconductivity, the superconducting transition temperature $T_c$ has remained until 1986 below 23.2 K, the $T_c$ of Nb$_3$Ge first discovered [2] in 1973. In the face of this gross failure to raise the $T_c$, nonconventional approaches [3] taking advantage of possible strong nonconventional superconducting mechanisms [4] have been proposed and tried. In September 1986, the situation changed drastically when Bednorz and Müller [5] reported the possible existence of percolative superconductivity in (La$_{1-x}$Ba$_x$)Cu$_{3-\delta}$ with $x = 0.2$ and 0.15 in the 30-K range. Subsequent magnetic studies [6–8] confirmed that high-temperature superconductivity indeed exists in this system. Takagi *et al.* further attributed the observed superconductivity in the La-Ba-Cu-O system to the K$_2$NiF$_4$ phase [9]. By the replacement of Ba with Sr [8,10,11], it is found that the La-Sr-Cu-O system of the K$_2$NiF$_4$ structure, in general, exhibits a higher $T_c$ and a sharper transition. A transition width [10] of 2 K and an onset [11] $T_c$ of 48.6 K were obtained at ambient pressure.

Pressure [8,12] was found to enhance the $T_c$ of the La-Ba-Cu-O system at a rate of greater than $10^{-3}$ K bar$^{-1}$ and to raise the onset $T_c$ to 57 K, with a "zero-resistance" state [13] reached at 40 K, the highest in any known superconductor until now. Pressure reduces the lattice parameter and enhances the Cu$^{+3}$/Cu$^{+2}$ ratio in the compounds. This unusually large pressure effect on $T_c$ has led to suggestions [8,12] that the high-temperature superconductivity in the La-Ba-Cu-O and La-Sr-Cu-O systems may be associated with interfacial effects arising from mixed phases; interfaces between the metal and insulator layers, or concentration fluctuations within the K$_2$NiF$_4$ phase; strong superconducting interactions due to the mixed valence states; or yet an unidentified phase. Furthermore, we found that when the superconducting transition width is reduced by making the compounds closer to the pure K$_2$NiF$_4$ phase, the onset $T_c$ is also reduced while the main transition near 37 K remains unchanged.

Extremely unstable phases displaying signals indicative of superconductivity in compounds consisting of phases in addition to or other than the $K_2NiF_4$ phase have been observed by us [8,14], up to 148 K, but only in four samples, and in China [15] at 70 K, in one sample. Therefore, we decided to investigate the multiple-phase Y-Ba-Cu-O compounds instead of the pure $K_2NiF_4$ phase, through simultaneous variation of the lattice parameters and mixed valence ratio of Cu ions by chemical means at ambient pressure.

The compounds investigated were prepared with nominal compositions represented by $(Y_{1-x}Ba_x)_2CuO_{4-\delta}$ with $x = 0.4$ through solid-state reaction of appropriate amounts of $Y_2O_3$, $BaCO_3$, and $CuO$ in a fashion similar to that previously described [8]. Bar samples of dimensions $1 \times 0.5 \times 4$ mm$^3$ were cut from the sintered cylinders. A four-lead technique was employed for the resistance $(R)$ measurements and an ac inductance bridge for the magnetic susceptibility $(\chi)$ determinations. The temperature was measured by means of Au+0.07% Fe-Chromel and Chromel-Alumel thermocouples in the absence of a magnetic field, and a carbon-glass thermometer in the presence of a field. The latter was calibrated against the former without a field. Magnetic fields up to 6 T were generated by a superconducting magnet.

The temperature dependence of $R$ determined in a simple liquid-nitrogen Dewar is shown in Fig. 1.1. $R$ initially drops almost linearly with temperature $T$. A deviation of $R$ from this $T$ dependence is evident at 93 K and a sharp drop starts at 92 K. A "zero-$R$" state is achieved at 80 K. The variation of $\chi$

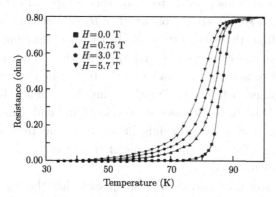

Fig. 1.1    Temperature dependence of resistance determined in a simple liquid-nitrogen Dewar.

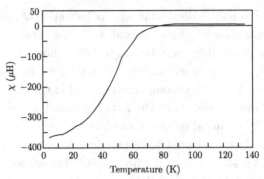

Fig. 1.2   Temperature dependence of magnetic susceptibility.

with $T$ is shown in Fig. 1.2. It is evident that a diamagnetic shift starts at 91 K and the size of the shift increases rapidly with further cooling. At 4.2 K, the diamagnetic signal corresponds to 24% of the superconducting signal of a Pb sample with similar dimensions. In a magnetic field, the $R$ drop is shifted toward lower $T$. At our maximum field of 5.7 T, the "zero-$R$" state remains at a $T$ as high as 40 K. Preliminary X-ray powder diffraction patterns show the existence of multiple phases uncharacteristic of the $K_2NiF_4$ structure in the samples. Detailed analyses and under way.

The above results demonstrate unambiguously that superconductivity occurs in the Y-Ba-Cu-O system with a transition between 80 and 93 K. We have determined the upper critical field $H_{c2}(T)$ resistively. If the positive curvature at very low fields is neglected, one gets a value of $dH_{c2}/dT$ near $T_c$ of 3 T/K or 1.3 T/K, depending on whether $H_{c2}(T_c)$ is taken at the 10% or the 50% drop from the normal-state $R$. In the weak-coupling limit, $H_{c2}(0)$ is thus estimated to be between 80 and 180 T in the Y-Ba-Cu-O system investigated. We believe that the value of $H_{c2}(0)$ can be further enhanced as the material is improved. The paramagnetic limiting field at 0 K for a sample with a $T_c \sim 90$ K is 165 T. Because of the porous and multiphase characteristics of the samples, it is therefore difficult to extract any reliable information about the density of states from the slope of $H_{c2}(T)$ at $T_c$ on the basis of the dirty-limit approximation.

On the basis of the existing data, it appears that the high-temperature superconductivity above 77 K reported here occurs only in compound systems consisting of a phase or phases in addition to or other than the $K_2NiF_4$ phase.

While it is tempting to attribute the superconductivity to possible nonconventional superconducting mechanisms as mentioned earlier, all present suggestions are considered to be tentative at best, especially in the absence of detailed structural information about the phases in the Y-Ba-Cu-O samples. However, we would like to point out here that the lattice parameters, the valence ratio, and the sample treatments all play a crucial role in achieving superconductivity above 77 K. The role of the different phases present in superconductivity is yet to be determined.

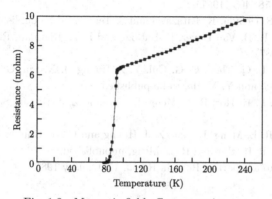

Fig. 1.3   Magnetic field effect on resistance.

The work at the University of Alabama at Huntsville is supported by NASA Grants No. NAG8-032 and No. NAGW-A12, and National Science Foundation Alabama EPSCoR Program Grant No. R11-8610669, and at the University of Houston by National Science Foundation Grant No. DMR 8616537, NASA Grants No. NAGW-977 and No. NAG8-051, and the Energy Laboratory of the University of Houston. Technical assistance from D. Campbell, A. Testa, and J. Bechtold is greatly appreciated.

[1]  C. W. Chu, U.S. Patent Application (12 January 1987).

[2]  J. R. Gavaler, Appl. Phys. Lett. **23**, 480 (1973); L. R. Testardi, J. H. Wernick and W. A. Royer, Solid State Commun. **15**, 1 (1974).

[3]  See, for example, C. W. Chu and M. K. Wu, in *High Pressure in Science and Technology*, edited by C. Homan, R. K. MacCrone and E. Whalley (North-Holland, New York, 1983), Vol. 1, p. 3; C. W. Chu, T. H. Lin, M. K.

Wu, P. H.Hor and X. C. Jin, in *Solid State Physics under Pressure*, edited by S. Minomura (Terra Scientific, Tokyo, 1985), p. 223.

[4] See, for example, J. Bardeen, in *Superconductivity in d- and f- Band Metals*, edited by D. Douglass (Plenum, New York, 1973), p. 1.

[5] J. G. Bednorz and K. A. Müller, Z. Phys. B **64**, 189 (1986).

[6] J. G. Bednorz, M. Takashige and K. A. Muller, to be published.

[7] S. Uchida, H. Takagi, K. Kitazawa and S. Tanaka, Jpn. J. Appl. Phys. **26**, L1 (1987).

[8] C. W. Chu, P. H. Hor, R. L. Meng, L. Gao, Z. J. Huang and Y. Q. Wang, Phys. Rev. Lett. **58**, 405 (1987).

[9] H. Takagi, S. Uchida, K. Kitazawa and S. Tanaka, to be published.

[10] R. J. Cava, R. B. Van Dover, B. Batlogg and E. A. Rietman, Phys. Rev. Lett. **58**, 408 (1987).

[11] Z. X. Zhao, L. Q. Chen, C. G. Cui, Y. Z. Huang, J. X. Liu, G. H. Chen, S. L. Li, S. Q. Guo and Y. Y. He, to be published.

[12] C. W. Chu, P. H. Hor, R. L. Meng, L. Gao and Z. J. Huang, Science **235**, 567 (1987).

[13] P. H. Hor, R. L. Meng, L. Gao, Z. J. Huang and C. W. Chu, to be published.

[14] C. W. Chu, P. H. Hor and R. L. Meng, unpublished.

[15] According to a report in Renmin Ribao, 17 January 1987.

# 2

# Superconductivity above Liquid Nitrogen Temperature in Ba-Y-Cu Oxides*

Z. X. Zhao, L. Q. Chen, Q. S. Yang, Y. Z. Huang, G. H. Chen, R. M. Tang, G. R. Liu, C. G. Cui, L. Chen, L. Z. Wang, S. Q. Guo, S. L. Li and J. Q. Bi

*Institute of Physics, Academia Sinica, Beijing*

## 2.1  Introduction

Bednorz and Müller [1] have discovered possible superconductivity around 35 K in Ba-La-Cu-O systems. Uchida and Takagi et al. [2,3] have observed Meissner effect and determined superconductivity in the system. Within a few months, several groups have reported their successful results [4–9]. Chu et al. [5] pushed the $T_c$ (onset) of Ba-La-Cu-O to 52 K under hydrostatic pressure. Cava et al. obtained nearly single superconducting phase with $T_c$ of 36.2 K and $\Delta T_c$ of 1.4 K. The present authors have achieved the result of $T_c$ (onset) for Sr-La-Cu-O and Ba-La-Cu-O determined resistively with $T_c$'s of 48.6 K, $\Delta T_c$ of 10 K and $T_c$'s of 46.3 K, $\Delta T_c$ of 7 K respectively, and observed that superconductivity might exist around 70 K. In this paper we report the experimental results of superconductivity in Ba-Y-Cu-O above liquid nitrogen temperature, breaking through the upper limit of $T_c$ about 40 K predicted theoretically by McMillan's theory. Based upon the understanding of $BaPb_{1-x}Bi_xO_3$ system, Rice and Sneddon [10] proposed a strong coupling electron-phonon model with local CDW instability. The model required the

* Reprinted from Z. X. Zhao, L. Q Chen, Q. S. Yang et al., Kexuetongbao **32**, 621 (1987).

electron-phonon interaction constant $\lambda$ to be close to 1. On the other hand, Chakraverty and Schlenker [11] gave a bipolaron theory where the value of $\lambda$ is required to be approximately 100. The superconducting La-Ba-Cu-O compound has a layer-perovskite structure with tetragonal (I4/mmm) symmetry, which means that this system has quasi-two-dimensional characteristics. Anderson [12] proposed a theory of the resonating-valence-bond state. Other theoretical works have also been put forward [13–15]. However, no satisfactory theory has yet been presented. More work is required to understand the superconducting mechanism of this kind of high $T_c$ superconductivity.

Uchida $et$ $al.$ and Takagi $et$ $al.$ have identified the superconducting phase as $(Ba_xLa_{1-x})_2CuO_4$ with $K_2NiF_4$ layer-perovskite structure. Most groups have accepted the above conclusion and obtained good results with $T_c$'s around 30–40 K. With this kind of composition it is easy to obtain the single superconducting phase. However, the present authors believe that although the sample with the nominal composition of $Ba(Sr)_{0.5}La_{4.5}Cu_5O_{5(3-y)}$ will be multi-phase in structure, it is beneficial to forming the high $T_c$ phase. In a nominal composition of $Ba_{0.5}La_{4.5}Cu_5O_{5(3-y)}$ for making the layer-perovskite stucture closer to the two-dimensional system, we have substituted La with Y to elevate the superconducting transition temperature. The experimental results show that the midpoint of the transition determined by resistivity is 92.8 K with transition width $\Delta T_c$ (10–90%) = 4 K, and that the onset of the transition is above 100 K. It can be observed clearly that the temperature dependence of resistance departs from linear behavior at about 110 K.

## 2.2   Experimental

Samples were prepared by solid state reaction with (s.p.) purity oxides of yttrium, barium and copper. These chemicals were dried at 130°C overnight before usage. The mixture was thoroughly ground and well mixed and heated in a tubular furnace for the synthesis with the following heat treatments in air: 400°C for 2 h, 800°C for 4 h followed by grinding and then sintering at 1000°C for 6 h and cooling in the furnace. Preliminary X-ray diffraction patterns show that the samples are of multiphase in structure.

Resistivity data were taken with a standard four-probe technique with 0.05-mm diameter Au wires pressed with In to a rectangular sample cut from the sintered pellet. The resistivity of the samples in normal state at 115 K was near 0.4 $\Omega \cdot$cm. The current density used for measurement was 0.04 A/cm$^2$ and the resolution of the voltage measurement was $2 \times 10^{-8}$ V.

The mutual-inductance method was adopted to measure the A. C. magnetic susceptibility with reference frequency of 314 Hz and time constant of 1 s. The output signal due to the transition was 8.2 $\mu$V. The resolution of the PAR lock-in amplifier was $2 \times 10^{-8}$ V.

The temperature was measured by using Pt thermometer calibrated with another Pt thermometer (which was calibrated by the Chinese Academy of Metrology) and checked against carbon-glass thermometer (which was calibrated by Lake Shore Company, model No. CGR-1-5000 serial No. C1383, calibration 158808417).

## 2.3   Results and Discussion

The resistance measurements show that the temperature coefficient of resistance is positive, and that no negative temperature coefficient appears in normal state, which we have observed in Sr(Ba)-La-Cu-O systems [4]. The data in Fig. 2.1 show that the resistance decreases linearly with temperature from 140 K to 115 K. At 110 K, the decrease of resistance obviously departs from linear behavior. This shows that the onset of the transition should be above 100 K. In this case, the measurement of resistance in magnetic field would be more effective to determine the onset of transition. This work is now in progress. The resistive transition is 4 K in width (10–90% of the resistance at 100 K) with the midpoint of the transition at 92.8 K. At 78.5 K, the resistance approaches zero in the region of sensitivity of the instrument. During the transition the resistance decreases five orders of magnitude.

The measured results of A. C. magnetic susceptibility show that the onset of diamagnetism is 93 K, wich corresponds to the midpoint of the transition measured resistively. It is shown in Fig. 2.1 that by comparing the signal of the sample with the voltage level of the standard Pb with the same volume, at the temperature of 77 K, the ratio of the superconducting phase should be about 13%, if the sample is a single-connected region.

The bydrostatic pressure method has been used to treat several Sr-La-Cu-O samples with $T_c$ about 40 K (these samples have very broad magnetic susceptibility transition from 40 K to 4.2 K, and the resistive transition width is about 10 K, i.e. from the beginning of the transition to zero resistance temperature). Under 45 kbar, the superconducting phase corresponding to $T_c$ of 30–40 K was increased but that corresponding to 15 K was decreased. The latter corresponds to the superconducting phase of a superstructure based on a cubic subcell with the parameter of 3.9 Å. When treated under 80 kbar, the superconductivity of the superstructure disappeared at 42 K. These results with be pubished elsewhere. Although the pressure treatment at room temperature did not elevate the critical temperature, the ratio of higher $T_c$ phase was obviously increased. As mentioned above, temperature dependence of resistance at 110 K started to depart obviously from linear behavior. This phenomenon shows that superconductivity may exist at 110 K. The above results suggest to us that using the high pressure treatment it might be possible to increase the superconducting phase even above 90 K.

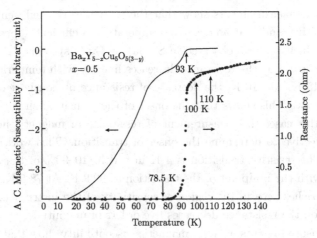

Fig. 2.1   Temperature dependence of resistivity (right scale) and A. C. magnetic susceptibility (left scale) of sample H-4 with $x = 0.5$. The resistive transition from the normal state to the superconducting state for $Ba_{0.5}$-$Y_{4.5}Cu_5O_{5(3-y)}$, clear onset at above 100 K, midpoint at 92.8 K and zero resistivity at 78.5 K with measurement current denfity 0.04 $A/cm^2$, the onset of diamagnetism at 93 K.

The authors would like to thank Prof. Jin Duo, Wang Changqing, Hu Boqing, Chen Jingran, Ren Shiyuan, Huang Xicheng, Cheng Jianbang, Ma Wenyi for their help in experiments.

[1]  J. G. Bednorz and K.A.Müller, Z. Phys. B. Condensed Matter **64**, 189 (1986).

[2]  S. Uchida, H. Takagi, K. Kitazawa and S. Tanaka, Japan J. Appl. Phys. Lett. (submitted).

[3]  H. Takagi, S. Uchida, K. Kitazawa and S. Tanaka, *ibid.* (submitted).

[4]  Z. X. Zhao, L. Q. Chen, C. G. Cui, Y. Z. Huang, J. X. Liu, G. H. Chen, S. L. Li, S. Q. Guo and Y. Y. He, Kexue Tongbao **32**, 8:522 (1987).

[5]  C. W. Chu, P. H. Hor, R. Meng, L. Gao, Z. H. Huang and Y. Q. Wang, Phys. Rev. Lett. **58**, 405 (1987).

[6]  R. J. Cava, R. B. Van Dover, B. Batlogg and E. A. Rietman, Phys. Rev. Lett. **58**, 408 (1987).

[7]  W. K. Kwok, G. W. Crabtree, D. G. Hinks, D. W. Capone, J. D. Jorgensen and K. Zhang (submitted); D. W. Capone, D. G. Hinks, J. D. Jorgensen and K. Zhang (submitted); J. D. Jorgensen, D. G. Hinks, D. W. Capone, K. Zhang, H. B. Schüttler and M. B. Brodsky (submitted).

[8]  D. K. Finnemore, R. N. Shelton, J. R. Clem, R. W. McCallum, H. C. Ku, R. E. McCarley, S. C. Chen, P. Klavins and V. Kogan (submitted).

[9]  C. Politis, J. Geerk, M. Dietrich and B. Obst, Z. Phys. B. Condensed Matter **66** (1987).

[10]  T. M. Rice and L. Sneddon, Phys. Rev. Lett **47**, 689 (1981); T. M. Rice, in *Superconductivity in Magnetic and Exotic Materials*, edited by T. Matsubara, A. Kotani (Springer, 1983).

[11]  B. K. Chakraverty and C. Schlenker, J. Phys. (Paris) Collog **37**, C4-353 (1976); B. K. Chakraverty, J. Phys. Lett. **40**, L99 (1979); B. K. Chakraverty, J. Phys. **42**, 1351 (1981).

[12]  P. W. Anderson, preprint.

[13]  H. B. Schüttler, J. D. Jorgensen, D. G. Hinks, D. W. Capone, and D. J. Scalapino (submitted).

[14]  C. S. Ting, D. Y. Xing and W. Y. Lai, preprint.

[15]  J. J. Yu, A. J. Freeman and J. H. Xu, submitted to PRL.

# 3

# Superconductivity above 130 K in the Hg-Ba-Ca-Cu-O System*

A. Schilling, M. Cantoni, J. D. Guo and H. R. Ott

*Laboratorium für Festkörperphysik, ETH Hönggerberg, 8093 Zürich, Switzerland*

The recent discovery [1] of superconductivity below a transition temperature ($T_c$) of 94 K in $HgBa_2CuO_{4+\delta}$ has extended the repertoire of high-$T_c$ superconductors containing copper oxide planes embedded in suitably structured (layered) materials. Previous experience with similar compounds containing bismuth and thallium instead of mercury suggested that even higher transition temperatures might be achieved in mercury-based compounds with more than one $CuO_2$ layer per unit cell. Here we provide support for this conjecture, with the discovery of superconductivity above 130 K in a material containing $HgBa_2Ca_2Cu_3O_{1+x}$ (with three $CuO_2$ layers per unit cell), $HgBa_2CaCu_2O_{6+x}$ (with two $CuO_2$ layers) and an ordered superstructure comprising a defined sequence of the unit cells of these phases. Both magnetic and resistivity measurements confirm a maximum transition temperature of ~133 K, distinctly higher than the previous established record value of 125–127 K observed in $Tl_2Ba_2Ca_2Cu_3O_{10}$ [2,3].

The structural similarity of $HgBa_2CuO_{4+\delta}$ (Hg-120 [1]) to a member of the thallium-containing family of copper oxides, $TlBa_2CuO_5$ (Tl-1201), suggests the existence of compounds with the general composition $HgBa_2Ca_{n-1}Cu_n$-$O_{2n+2+\delta}$. The transition temperatures of the thallium-containing analogues, $TlBa_2Ca_{n-1}Cu_nO_{2n+3}$, range from < 10 K ($n = 1$ [4]) to ~110 K ($n = 3$ [5]). In this sense, transition temperatures exceeding 100 K may be expected also

---

* Reprinted from A. Schilling, M. Cantoni, J. D. Guo and H. R. Ott, Nature **363**, 56 (1993).

in the Hg-Ba-Ca-Cu-O (HBCCO) system. Although the successful synthesis of $HgBa_2RCu_2O_{6+x}$ (Hg-1212) with R being (Eu, Ca) has been reported, no superconductivity was found in that system [6].

We prepared the samples following the procedure described in [1] for Hg-1201. A precursor material with the nomial composition $Ba_2CaCu_2O_5$ was obtained from a well ground mixture of the respective metal nitrates, sintered at 900°C in $O_2$. After regrinding and mixing with powdered HgO, the pressed pellets were sealed in evacuated quartz tubes. These tubes were placed horizontally in tight steel containers and held at 800°C for 5 hours. On opening the containers, we found that the quartz tubes were broken. It was not possible to reconstruct at which stage of the heating, cooling or opening procedure this happened. Some of the pellets were finally annealed for 5 hours at 300°C in flowing oxygen. During the preparation and the characterization of the samples, all possible measures were taken to avoid any contamination with toxic mercury of mercury-containing compounds.

After annealing, the resulting black material was characterized by X-ray diffraction using the Guinier technique, by energy-dispersive X-ray spectrometry (EDS), and by selected-area electron-diffraction techniques (SAED) and high-resolution transmission electron microscopy (HRTEM). The EDX analysis showed that the samples are composites of isolated grains of $BaCuO_2$ ($\sim$30%), CuO ($\sim$30%), an unidentified oxide containing Hg, Ca and Cu ($\sim$15%), an oxide with Ca and Cu ($\sim$5%), and $\sim$5% impurities with unspecified composition. About 15% of the total sample volume consisted of plate-like grains containing Hg, Ba, Ca and Cu. Some of these were investigated in detail by SAED and HRTEM techniques on a Phillips CM 30-ST transmission electron microscope. Both techniques showed clearly that these identified grains consist mostly of pure $HgBa_2Ca_2Cu_3O_{8+x}$ (Hg-1223), disordered mixtures of Hg-1223 and Hg-1212, and a periodic stacking sequence of the latter unit cells. We found no grains or intergrowths associated with the Hg-1201 structure. As the volume fraction of the phases of interest is fairly small, we could not measure the lattice parameters precisely with X-ray Guinier technique. Nevertheless, from the SAED patterns, we deduce the lattice constants $c = 12.7(2)$ Å and $c = 16.1(3)$ Å for the tetragonal Hg-1212 and Hg-1223 units, respectively, and $a = 3.93(7)$ Å, valid for both types of compounds. The results for Hg-1212 are in good agreement with the values obtained in

[6]. Figure 3.1 is a representative HRTEM image showing, as an example, a stacking containing both Hg-1212 and Hg-1223 layers. The stacking sequence 1223/1223/1212/1212/1223/1212 with a supercell $c$-axis $c \approx 86.4$ Å extends beyond 2000 Å, thus qualifying this superstructure as a proper phase. HRTEM images as well as SAED patterns gave no evidence for the presence of HgO-double layers.

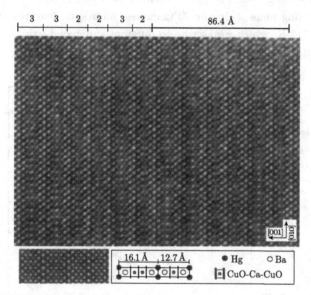

Fig. 3.1   HRTEM image of a grain in [100] orientation, containing layers of Hg-1212 and Hg-1223. Here, they are stacked in a periodic sequence forming a supercell with $c \approx 86.4$ Å (see text). A contrast simulation ($c_s = 1.1$ mm, $E = 300$ keV, defocus $-870$ Å, specimen thickness 23 Å) is inserted. The stacking sequence in terms of the number of Cu-O planes and an enlarged schematic drawing of the involved unit cells are included.

We measured the magnetic susceptibility of the specimens using a SQUID-magnetometer (Quantum Design). Figure 3.2 shows the result obtained for an oxygen-annealed sample. In an external field $H = 27$ Oe, the zero-field cooling susceptibility (ZFC) amounts to ~100% of $1/4\pi$ at temperature $T = 6$ K, indicating complete magnetic screening. For this estimate, we assumed an average density $\rho = 6$ g·cm$^{-3}$. The field-cooling (FC) susceptibility reaches ~10% of the maximum possible value. This value represents a lower-bound value for the true superconducting volume fraction in the sample, indicating the bulk nature of superconductivity. The onset temperature of diamagnetism is $T_c = 133.5$ K, seen both in FC and ZFC experiments (see Fig. 3.2, inset). The FC susceptibility reaches ~60% of its full lowtemperature value at 125 K, strongly indicating that the phase with $T_c \approx 133.5$ K dominates all other superconducting phases. In the ZFC curves, additional features are seen at 126 K and 112 K, which we ascribe to different superconducting phases with lower transition temperatures.

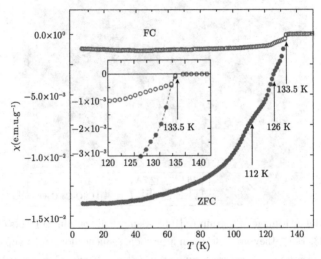

Fig. 3.2    Zero-field cooling (ZFC) and field cooling (FC) susceptibilities $\chi(T)$ of one of the investigated oxygen-annealed HBCCO samples, measured in $H = 27$ Oe. The ZFC curve indicates the presence of several different superconducting phases.

The resistivity $R$ as a function of temperature T of an annealed sample is shown in Fig. 3.3. At $T = 132.5$ K, $R(T)$ drops sharply with a maximum in the differential d$R$/d$T$, and reaches zero at $T = 95$ K within the resolution of the four-probe a.c.-resistance bridge used. This temperature is still considerably higher than the zero-resistance temperature $T = 35$ K, reported for Hg-1201 [1]. The final oxygen treatment was very effective in increasing the critical temperature; the as-sintered samples showed a maximum $T_c$ of only $\sim 117$ K.

At present we cannot relate the different superconducting phases to crystallographic phases. There is no unambiguous proof that the occurrence of superconductivity in our samples stems from the $HgBa_2Ca_{n-1}Cu_nO_{2n+2+\delta}$ phases. In analogy with the thallium- and bismuth-based copper oxides [5], however, we suggest that in the HBCCO system $T_c$ also increases with the number of Cu-O planes per unit cell, and conclude that Hg-1223 is responsible for superconductivity at $\sim$133 K. This would be consistent with the large relative superconducting volume fraction at 125 K, in view of the dominance of Hg-1223 observed in the grains investigated microscopically.

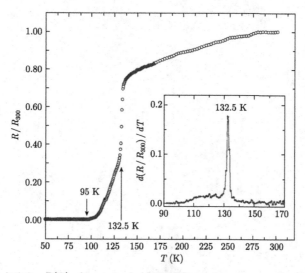

Fig. 3.3    Resistivity $R(T)$ of an annealed HBCCO specimen, normalized with respect to the resistance value $R(300) \approx 0.10\ \Omega$. The inset displays the temperature derivative $dR/dT$ to show the maximum resistivity drop at $T \approx 132.5$ K. Zero resistance is attained at $T = 95$ K.

We thank S. Ritsch for his help in the structural characterization. This work was supported in part by the Schweizerische Nationalfonds zur Förderung der wissenschaftlichen Forschung.

[1]  S. N. Putilin, E. V. Antipov, O. Chmaissem and M. Marezio, Nature **362**, 226 (1993).

[2]  T. Kaneko, H. Yamauchi and S. Tanaka, Physica C **178**, 377 (1991).

[3]  S. S. P. Parkin *et al.*, Phys. Rev. Lett. **60**, 2539 (1988).

[4]  I. K. Gopalakrishnan, J. V. Yakhmi and R. M. Iyer, Physica C **175**, 183 (1991).

[5]  S. S. P. Parkin *et al.*, Phys. Rev. Lett. **61**, 750 (1988).

[6]  S. N. Putilin, I. Bryntse and E. V. Antipov, Mat. Res. Bull. **26**, 1299 (1991).

# 4

# Pairing Symmetry and Flux Quantization in a Tricrystal Superconducting Ring of YBa$_2$Cu$_3$O$_{7-\delta}$*

C. C. Tsuei, J. R. Kirtley, C. C. Chi, L. S. Yu-Jahnes, A. Gupta, T. Shaw, J. Z. Sun and M. B. Ketchen

*IBM Thomas J. Watson Research Center, P.O. Box 218, Yorktown Heights, New York 10598*

We have used the concept of flux quantization in superconducting YBa$_2$Cu$_3$O$_{7-\delta}$ rings with 0, 2, and 3 grain-boundary Josephson junctions to test the pairing symmetry in high-$T_c$ superconductors. The magnetic flux threading these rings at 4.2 K is measured by employing a scanning superconducting quantum interference device microscope. Spontaneous magnetization of a half magnetic flux quantum, $\Phi_0/2 = h/4e$ has been observed in the 3-junction ring, but not in the 2-junction rings. These results are consistent with $d$-wave pairing symmetry.

An unambiguous determination of the order parameter symmetry is crucial to understanding the mechanism responsible for high-temperature superconductivity in the cuprates. For example, a $d_{x^2-y^2}$ symmetry [$\delta(\hat{k}) \sim k_x^2 - k_y^2 \sim \cos 2\theta$] in the pairing wave function will lend strong support for pairing mediated by antiferromagnetic spin fluctuations [1]. Theories such as the interlayer coupling model [2] will be supported if $s$-wave superconductivity can be demonstrated. On the other hand, the van Hove model [3] is compatible with either symmetry. Recently, there have been numerous experiments [4–6]

---

dealing with various aspects of pairing symmetry in high-$T_c$ superconductors. Unfortunately, the results are ambiguous. The reports in favor of $d$-wave pairing are roughly equal in number to those supporting $s$-wave symmetry. This unsettling situation partially stems from the often indirect nature of the experimental evidence, as well as experimental issues such as sample quality, impurity scattering, twinning, and trapped magnetic flux. For details, the reader is referred to Ref. [4].

In this Letter an experiment based on flux quantization of a three-grain ring of YBa$_2$Cu$_3$O$_{7-\delta}$ is proposed and applied to test the symmetry of the pair state. The symmetry of a pair wave function can best be probed at the junction interface as the Cooper pairs tunnel across a Josephson junction or weak link [7]. The sign of the Josephson current of a junction between two $d$-wave superconductors depends on the relative orientation of their order parameters with respect to the junction interface. As shown recently by Sigrist and Rice [8], the supercurrent $I_s^{ij}$ can be expressed by

$$I^{ij} = \left( A^{ij} \cos 2\theta_i \cos 2\theta_j \right) \sin \Delta\phi_{ij} = I_c^{ij} \sin \Delta\phi_{ij}, \qquad (4.1)$$

where $A^{ij}$ is a constant characteristic of junction $ij$, and $\theta_i$ and $\theta_j$, are angles of the crystallographic axes (or equivalently wave vectors $k_x$ and $k_y$) with respect to a junction interface (e.g., a grain boundary) between superconductors $i$ and $j$.

In the case of $s$-wave symmetry, the sign of $I_s^{ij}$ is a constant of $\theta_i, \theta_j$, but its magnitude can vary due to gap anisotropy. For the case of a single superconducting ring with one Josephson junction, Sigrist and Rice showed that, based on free energy considerations, the ground state of a ring with one $\pi$ junction (i.e., $I_c^{ij} < 0$) has a spontaneous magnetization if the critical current is sufficiently large. The magnetic flux threading through such a $\pi$ ring is exactly half of the flux quantum ($\Phi_0/2 = h/4e = 1.035 \times 10^{-7}$ G $\cdot$ cm$^2$) when the external field $H_{\mathrm{ext}} = 0$ and the condition $L|I_c| > \Phi_0$, where $L$ is the self-inductance of the ring, is satisfied. In the case of a multiple-junction ring, Sigrist and Rice [8] suggested that a ring of odd-number $\pi$-junctions will also exhibit $\Phi_0/2$ spontaneous magnetization. In the following, we show that this is indeed true. In reference to the odd-number $\pi$-junction ring in Fig. 4.1 and Eq. (4.1), the flux quantization of a superconducting ring can be written as follows:

$$\Phi_{\text{ext}} + I_s L + \frac{\Phi_0}{2\pi} \sum_{ij} \Delta\phi_{ij} = n\Phi_0, \tag{4.2}$$

where $I_s$, the current circulating in the ring, is given by

$$I_s = I_c^{12} \sin \Delta\phi_{12} = I_c^{23} \sin \Delta\phi_{23} = I_c^{ij} \sin \Delta\phi_{ij}. \tag{4.3}$$

Equations (4.2) and (4.3) are valid for both s-wave and d-wave rings. In the case of a ring with an odd number of $\pi$ junctions, it is sufficient to consider the case where only one critical current is negative (say $I_c^{12} = -|I_c^{12}|$). Then $I_s = |I_c^{12}| \sin(-\pi + \Delta\phi_{12})$. For zero external field, $\Phi_{\text{ext}} = 0, n = 0$, and $|I_c^{12}|L \gg \Phi_0, I_c^{ij}L \gg \Phi_0$, the flux quantization condition Eq. (4.2) will lead to the following expression for the circulating current:

$$I_s = \frac{\pi}{2\pi(L/\Phi_0) + 1/|I_c^{12}| + 1/I_c^{23} + \cdots} \simeq \frac{\Phi_0}{2L}. \tag{4.4}$$

Equation (4.4) is essentially a restatement of the spontaneous magnetization of an odd-number $\pi$-junction ring.

It should be pointed out that the Sigrist-Rice formula, Eq. (4.1), is based on an implicit assumption that the junction interface is perfectly smooth and without any disorder. In reality, the electron wave vector orthogonal to the junction face can be significantly distorted by interface roughness, impurities, strain, oxygen deficiency, etc.

To model the disorder effect, one can consider the consequence of an angular deviation $\Delta\theta$ from the perfect junction interface. The results of a straightforward calculation can be expressed as

$$I_s = \frac{1}{2} A^{ij} \{\cos 2(\theta_i + \theta_j) + B^{ij}(\alpha) \cos 2(\theta_i - \theta_j)\} \sin \Delta\phi_{ij}, \tag{4.5}$$

where $B^{ij}(\alpha)$ is a function of the angular distribution probability $P(\theta)$ of $\theta_i$ and $\theta_j$, $\int_{-\alpha}^{\alpha} P(\theta)d\theta = 1$, and $\alpha$ is a measure of the angular variation of $\theta$. As a consequence of the fourfold symmetry in the Cu-O square lattice in the $CuO_2$ planes, the maximum disorder at the grain boundary corresponds to $\alpha = \pi/4$. This leads to $B^{ij}(\pi/4) = 0$, and a maximum-disorder formula for the circulating current:

$$I_s^{ij} = I_c^{ij} \cos 2(\theta_i + \theta_j) \sin \Delta\phi_{ij}. \tag{4.6}$$

Given Eq. (4.1) and Eq. (4.6), one can design a $d$-wave pairing symmetry test for cuprate superconductors that is valid for all cases. The basic idea of our experiment is to fabricate a trigrain boundary-junction ring of high-$T_c$, superconductor such as $YBa_2Cu_3O_{7-\delta}$. A scanning SQUID (superconducting quantum interference device) [9] microscope is then used to image the magnetic flux threading through the superconducting trigrain ring to search for the $\Phi_0/2$ spontaneous magnetization.

The three-junction rings used in this work are fabricated from epitaxial films of $YBa_2Cu_3O_{7-\delta}$ deposited on a (100) tricrystal $SrTiO_3$ substrate [10] (Fig. 4.1) by using standard pulsed laser deposition. The $YBa_2Cu_3O_{7-\delta}$ film is $c$-axis oriented with $c = 11.685$ Å and a zero resistance transition temperature of 90.7 K. The misorientation across each grain boundary in the tricrystal substrate (as indicated in Fig. 4.1) was determined from backscattered Kikuchi maps recorded from each crystal using the electron beam probe of a scanning electron microscope. For each boundary the misorientation angle was within $4°$ of the design misorientation angle of $30°$. The rotation axes of the boundaries was close to the normal to the plane of the sample, showing that the boundaries are primarily of tilt type character.

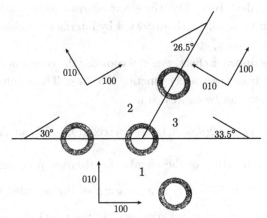

Fig. 4.1   Schematic diagram for the tricrystal (100) $SrTiO_3$ substrate, with four epitaxial $YBa_2Cu_3O_{7-\delta}$ rings.

In this experiment, four rings (inner diameter 48 μm, width 10 μm) are patterned using a standard photolithographic process. To test the quality of the individual grain boundary junctions across each grain boundary, bridges 25 μm in length and 10 μm in width across each grain boundary are prepared on bicrystal substrates that were cut off from the tricrystal substrate. The epitaxial $YBa_2Cu_3O_{7-\delta}$ films for these junctions were laser deposited in the same run and in close proximity to the tricrystal substrate for the four rings. The values of $I_c^{ij}(J_c^{ij})$ for these three test junctions agree within 20%. $I_c^{12}(J_c^{12})$, for example, is found to be 1.8 mA ($1.5 \times 10^5$ A/cm$^2$). The resistance as a function of temperature, $R(T)$, for the bridge across a grain boundary has a small shoulder below the $T_c$ ($= 90.7$ K) characteristic of a grain boundary weak link. The $IV$ curve ($T = 4.2$ K) exhibits a typical resistively shunted Josephson junction characteristic. From the $I_c^{ij}$ value and the estimated self-inductance of rings ($L \simeq 100$ pH) one finds that the $LI_c^{ij}$ product is about $100\,\Phi_0$, easily satisfying the condition $LI_c^{ij} \gg \Phi_0$ of Eq. (4.4). Therefore, a spontaneous magnetization of $\Phi_0/2$ at $\Phi_{ext} \sim 0$ should be observable in our 3-junction ring.

Figure 4.2 shows a scanning SQUID microscope [9] image of the four rings in our experiment. This image was obtained by scanning the pickup loop of a SQUID [shown schematically in Fig. 4.3(a)] relative to the sample at 4.2 K. The loop center was 10μm from the sharpened tip of its substrate, which was in direct contact with the sample, and rotated approximately 20° away from the sample plane, with the leads oriented towards the top of the image. The sample was mounted on a flexible cantilever, so that the tip remained in contact while scanning. The SQUID was operated in flux-locked mode, with noise approximately $2 \times 10^{-6}\,\Phi_0/\text{Hz}^{1/2}$. The ratio of the mutual inductance between loop and ring to the self-inductance of the ring is about 0.02, so that the effect of the SQUID flux coupling back into the ring should be small. Our interpretation of this image (run 12 in Fig. 4.3) is that the upper-right and left 2-junction rings, and the lower-right 0-junction ring, have no flux threading them, while the center 3-junction ring has 1/2 of the flux quantum $h/2e$ threading it. The outer control rings are visible through mutual inductance coupling between the rings and the SQUID loop. We determine the amount of flux in the central ring as follows.

The mutual inducantance $M(\rho)$ between a pickup loop tilted at an angle $\theta$ from the sample ($x-y$) plane in the $x$-$z$ plane, and a circular wire of radius $R$ at the origin, is given by

$$M(\boldsymbol{\rho}) = \frac{\mu_0 R}{4\pi} \int d^2x \int_0^{2\pi} d\phi \frac{\cos\theta(R - y\sin\phi - x\cos\phi) - \sin\theta(z\cos\phi)}{(x^2 + y^2 + z^2 + R^2 - 2xR\cos\phi - 2yR\sin\phi)^{3/2}},$$

(4.7)

where the integral $d^2x$ is over the plane of the pickup loop, and the vector $\boldsymbol{\rho}$ specifies the displacement of the pickup loop with respect to the ring in the $x$-$y$ plane. We calculate $M(0) = 2.4$ pH for the as fabricated tip centered above a 29 μm radius ring, at a tilt angle of 20°. A given flux $\Phi$ threading a superconducting ring with self-inductance $L$ induces a circulating current $I_r = \Phi/L$ around the ring, which in turn induces a flux $\Phi_s(\boldsymbol{\rho}) = M(\boldsymbol{\rho})\Phi/L$ in the pickup (sensor) loop. We calculate the inductance of our rings to be $99 \pm 5$ pH.

The solid lines in the bottom part of Fig. 4.2 are model calculations for the cross sections indicated by the contrasting lines in the image, assuming

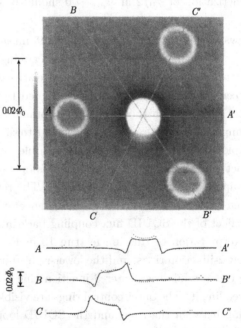

Fig. 4.2   Scanning SQUID microscope image of the four superconducting rings in Fig. 4.1, cooled in a field < 5 mG. The dots are cross sections through the data as indicated by the contrasting lines. The solid lines are calculations assuming the 3-junction ring has $\Phi_0/2$ flux threading it.

$\Phi = \Phi_0/2 = h/4e$ in the 3- junction ring. The asymmetry in the images results from the tilt of the pickup loop, as well as the asymmetric pickup area from the unshielded section of the leads. Using $\Phi_0/2$ for the flux in the 3-junction ring results in much better agreement than would be obtained using $\Phi_0$.

Our value for $M(0)$ was checked by positioning the pickup loop in the centers of the rings and measuring the SQUID output vs field characteristic. Representative results for the 3-junction ring are shown in Fig. 4.3(a). In this figure a linear background, measured by placing the loop over the center of the 0-junction control ring, has been subtracted out. The upper inset in this

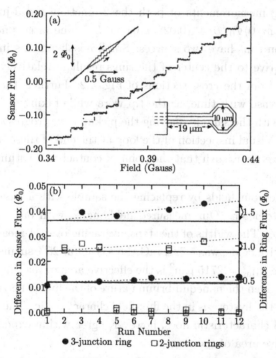

Fig. 4.3   (a) The measured SQUID flux vs field characteristic with the SQUID pickup coil centered on the 3-junction ring. The upper left inset shows the flux-field characteristic over a larger field range. The lower right inset shows the geometry of the SQUID pickup loop, with 1.2 μm linewidths and spacings. (b) The result of 12 separate cooldowns of the sample in nominal zero field. The solid lines are our calculations for the as-fabricated tip, the dashed lines include a correction for tip wear.

figure shows the sensor flux vs field characteristic over a larger field range. At low fields stepwise admission of flux into the ring leads to a staircase pattern, with progressively smaller heights and widths to the steps, until over a small intermediate field range, shown for increasing field in the main part of Fig. 4.3(a), single flux quanta are admitted. At larger fields the steps disappear and the SQUID flux vs field characteristic slowly oscillates about a mean line. The heights of the single flux-quantum steps in the intermediate field region, derived by fitting the data with a linear staircase (dashed line) are $\Delta \Phi_s = 0.0237 \Phi_0$. This is in good agreement with our calculated value of $\Delta \Phi_s = M(0) \Phi_0 / L = (0.024 \pm 0.003) \Phi_0$. Twelve repetitions of this measurement, including measurements of both the 2-junction and 3-junction rings, gave values of $M(0) \Phi_0 / L = (0.028 \pm 0.005) \Phi_0$. The large uncertainties in these calibration runs have two sources: Small misalignments in the position of the loop relative to the center of the rings result in relatively large errors, as can be seen from the cross sections of Fig. 4.2. Further, the step heights on average increase with time, as the tip wore while taking $\simeq 100$ images in direct contact with the sample, moving the pickup loop progressively closer to the ring plane. Visual inspection of the loop at the end of these measurements showed extensive wear, such that the point of contact was within 2 μm of the pickup loop edge.

We calibrated our fields by replacing the sample with a large pickup area SQUID magnetometer. Our measured fields agree with our calculations to within about 3%. The widths of the steps, averaging over 11 measurements for increasing positive fields, was 5.7±1 mG. This is about 25% smaller than $\Delta B = \Phi_0 / A_{\text{ring}}$, where $A_{\text{ring}} = 2715$ μm$^2$ is the effective area of the rings. This is not too surprising, given the nonequilibrium nature of the flux penetration process, as indicated by the hysteresis in the flux-field characteristic. The average slope of the flux-field characteristic does, however, agree within experimental error with the effective area of the rings.

Figure 4.3(b) summarizes the results from 12 cooldowns of the sample. We plot the absolute value of the difference between the SQUID loop flux in the centers of the 2-junction or 3-junction rings, and the 0-junction control ring. Since each point was taken from a full image, we could judge the center of the rings with accuracy, and our data scatter is much smaller than in the calibration runs. The solid lines are the expected values for the flux difference, calculated as described above. In all of our measurements $\Delta\Phi$ always fell close to $(N + 1/2)h/2e$ for the 3-junction ring, and close to $Nh/2e$ for the 2-junction rings ($N$ an integer). However, there is clearly some drift to the data, which we associate with tip wear. A fit to the eight $\Phi_0/2$ points in Fig. 4.3(b), assuming exactly $h/4e$ flux threads the 3-junction rings, implies that the mutual inductance $M(0) = 2.4$ pH for the as fabricated tip, and increases to 2.9 pH at the end of the series. For comparison, our calculations give 2.4 pH for the center of the loop 10 μm from the tip end, and 2.7 pH for the tip end just at the edge of the pickup loop. The dashed lines, including this correction to the mutual inductance, agree remarkably well with the data.

One can infer the inductance of the rings from the high-field asymptotic difference between the flux in a ring with junctions and the flux in the ring without junctions (Fig. 4.3). At the end of our measurement series this (SQUID) flux was $(4.0 \pm 0.2)\Phi_0/G$ of applied field. Using $L = A_{\mathrm{ring}}M(0)/(d\Phi_s/dB)$, and a mutual inductance of 2.9 pH, gives $L = 95 \pm 5$ pH, in good agreement with our calculation of $L = 99 \pm 5$ pH. For our present interpretation to be incorrect, (1) our calculations of the inductances would have to be incorrect by a factor of 2, (2) the flux steps in our calibration runs would have to go immediately from 2 flux quanta/step to no steps, and (3) the flux in the 3-junction ring would have to always be odd-integer values different from the control ring, while the 2-junction rings would always have to have even integer differences from the control. This seems extremely unlikely.

Although our results are consistent with $d$-wave pairing symmetry, this experiment alone cannot rule out even parity states with an order parameter symmetry varying as $\Delta \sim \cos(4\theta)$ [11]. It has also been proposed that magnetic spin-flip scattering or correlation effects at the grain boundary can induce a $\pi$ shift in the superconducting order parameter phase at each grain boundary [12].

In summary, we have directly observed for the first time spontaneous magnetization of $\pm\Phi_0/2$ in a 3-junction ring of epitaxial $YBa_2Cu_3O_{7-\delta}$. This observation supports $d$-wave pairing symmetry.

The scanning SQUID microscope used for this work was developed under the auspices of the Consortium for Superconducting Electronics. The authors wish to thank D. H. Lee for many stimulating discussions, in particular on the effect of disorder on the grain-boundary characteristics. They are thankful to

W. Gallagher, D. M. Newns, P. Chaudhari, and other colleagues for useful
discussions. The technical assistance of G. Trafas, M. Cali, and J. Hurd is
greatly appreciated.

[1]  For example, N. E. Bickers, D. J. Scalapino and S.R. White, Phys. Rev. Lett.
     **62**, 961 (1989); P. Monthoux, A. V. Balatsky and D. Pines, Phys. Rev. B **46**,
     14803 (1992).
[2]  S. Chakravarty *et al.*, Science **251**, 337 (1993).
[3]  R. S. Markiewicz, J. Phys. Condens. Matter **2**, 665 (1990); D. M. Newns *et
     al.*, Comments Condens. Matter Phys. **15**, 273 (1992); C.C. Tsuei *et al.*, Phys.
     Rev. Lett. **69**, 2134 (1992), and references therein.
[4]  For a brief review, see an article by B.G. Levi, in Phys. Today **46**, No. 5, 17
     (1993), and references therein; also the letters in Phys. Today **47**, No. 2, 11
     (1994).
[5]  D. A. Wollman *et al.*, Phys. Rev. Lett. **71**, 2134 (1993).
[6]  P. Chaudhari and Shawn-Yu Lin, Phys. Rev. Lett. **72**, 1084 (1994).
[7]  V. B. Geshkenbein, A. I. Larkin and A. Barone, Phys. Rev. B **36**, 235 (1987).
[8]  Manfred Sigrist and T.M. Rice, J. Phys. Soc. Jpn. **61**, 4283 (1992).
[9]  F.P. Rogers, BS/MS thesis, EICS Department, MIT (1983); R. C. Black *et al.*,
     Appl. Phys. Lett. **62**, 2128 (1993); L. N. Vu, M. S. Wistrom and D. J. van
     Harlingen, Appl. Phys. Lett. **63**, 1693 (1993).
[10] Tricrystal designed by the authors and manufactured by Shinkosa Co., Tokyo.
[11] C. M. Varma (private communication).
[12] L. N. Bulaevski, V. V. Kuzii and A. A. Sobyanin, JETP Lett. **25**, 290 (1977);
     B.I. Spivak and S. Kivelson, Phys. Rev. B **43**, 3740 (1991).

# 5

# NMR Study of Local Hole Distribution, Spin Fluctuation and Superconductivity in Tl$_2$Ba$_2$Ca$_2$Cu$_3$O$_{10}$*

G. Q. Zheng[1], Y. Kitaoka[1], K. Asayama[1], K. Hamada[2], H. Yamauchi[2], S. Tanaka[2]

[1] Department of Material Physics, Osaka University, Osaka 560, Japan
[2] Superconductivity Research Laboratory, ISTEC, Koto-Ku, Tokyo 135, Japan

$^{63}$Cu, $^{17}$O and $^{205}$Tl NMR have been performed in the high-$T_c$ superconductor Tl$_2$Ba$_2$Ca$_2$Cu$_3$O$_{10}$ whose $T_c$ (max) is 127 K. The hole densities at Cu and oxygen sites in the CuO$_2$ plane have been extracted from the nuclear quadrupole frequency $\nu_Q$. The striking feature is that the Cu holes are significantly transferred to oxygen site due to strong hybridization between Cu and oxygen. From an analysis of $T_1$ and $T_{2G}$, it has been found that the spectral weight of the spin fluctuation is transferred to higher energy compared to YBa$_2$Cu$_3$O$_7$, while the magnetic correlation length $\xi$ does not differ much. Thus, it is suggested that the higher $T_c$ is due to higher characteristic energy of spin fluctuations, i.e. the superconductivity is spin fluctuation mediated. The superconducting properties are consistently explained by a $d$-wave superconductivity model with a finite density of states (DOS) at the Fermi level. We show that the disorder of the Ca/TlO layer caused by the partial inter-substitution of Tl and Ca is responsible for the potential scattering to produce such a DOS. It is found that if such a potential scattering were absent, $T_c$ would go up to 132 K which is quite close to the record $T_c$ realized in the Hg based compound.

---

\* Reprinted from G. Q. Zheng, Y. Kitaoka, K. Asayama *et al.*, Physica C **260**, 197 (1996).

## 5.1   Introduction

Vast studies have revealed that many of the normal-state properties, including the spin and charge dynamics, such as the $T$ linear resistivity, the unusual Cu nuclear relaxation rate, are dominated by the antiferromagnetic (AF) Cu spin fluctuations [1]. Whether the superconductivity is also driven by such spin fluctuation or by other sources such as charge fluctuations, however, is still controversial [2–6]. The central issues, we believe, are an extract of the conditions which is favorable for $T_c$, of the microscopic hole distribution and the spin dynamics, and the identification of the symmetry of the superconducting order parameter, both of which should help identifying the mechanism for high-$T_c$ superconductivity.

One of the approaches to the problems that we have to settle may be a comparative study of systems with different doping conditions, which are available by applying high pressure, substituting heterovalent elements or adopting one with different favorable structure. From the hole redistribution between Cu and oxygen under pressure as clarified by NMR/NQR, it has been proposed that not only the doping rate, but also the hole distribution between Cu and O are important parameters in determining $T_c$ [7–9]. Due to the covalency effect between Cu and oxygen, the electronic configuration of Cu is not always of $3d^9$, with one hole in the $3d$ orbit. The transferring of Cu hole into oxygen, i.e., the hybridization between Cu$3d$ and O$2p$ orbits, results in the suppression of AF spin fluctuation of low-energy components [8]. High pressure is found to suppress the spin fluctuation of low energy in $La_{1.85}Sr_{0.15}CuO_4$ while it increases the density of states in $YBa_2Cu_4O_8$, and thereby increases $T_c$ [9].

$Tl_2Ba_2Ca_2Cu_3O_{10}$ (Tl2223 for short) with $T_c(\mathrm{max}) = 127$ K [10,11] having been the highest $T_c$ until the recent discovery of $HgBa_2Ca_2Cu_3O_8$ (Hg1223) [12], provides a good opportunity to extract the conditions favorable for high-$T_c$. There are two distinct $CuO_2$ planes in this system; the Cu(1) has a four-fold coordination (square $CuO_2$ plane), while the Cu(2) has a five-fold coordination (pyramidal plane) (Fig. 5.1). A part of the normal-state results on the pyramidal plane has already been published [13]. We present in this paper a comprehensive study of the $^{63}$Cu, $^{17}$O and $^{205}$Tl NMR in a sample of $Tl_{1.7}Ba_2Ca_{2.3}CuO_{10}$ with $T_c = 115$ K. We found, from analyzing $\nu_Q$, that the Cu hole is only 0.7 which is the smallest in the known compounds, indicating a strong covalency effect between Cu and O. A comparison with other compounds shows that a strong covalency is favorable for high $T_c$. An analysis of $T_1$ and $T_{2G}$ based on the model of Millis, Monien and Pines (MMP) [14] shows that the spectral weight of the spin fluctuation is transferred to higher energy compared to YBa$_2$Cu$_3$O$_7$, while the magnetic correlation length $\xi$ does not differ much. The product of the characteristic energy $\Gamma_Q$ and $\xi^2$ is larger by 40% than that in YBa$_2$Cu$_3$O$_7$, well accounting for the higher $T_c$ in this system, in the mechanism of spin fluctuation induced superconductivity.

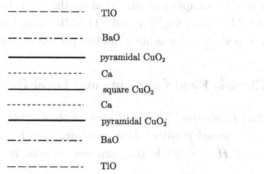

Fig. 5.1    Block diagram of the crystal structure of Tl$_2$Ba$_2$Ca$_2$Cu$_3$O$_{10}$.

In the superconducting state, the $^{63}(1/T_1T)$ and the Knight shift $K$ are compatible with the $d$-wave superconducting model with a finite density of states (DOS) at the Fermi surface caused by potential scattering. It is found that the disorder of the Ca layer is the origin of potential scattering.

This paper is organized as follows. In Section 5.2, we give a brief description of the experimental procedure. In Section 5.3, we extract the charge densities at Cu and oxygen sites by analyzing $\nu_Q$. In Sections 5.4 and 5.5, the results of $T_1$ and $T_{2G}$ are presented and the character of the spin fluctuation is extracted based on the model of MMP. The superconducting properties are discussed in Sections 5.6 and 5.7. We summarize and conclude in Section 5.8.

## 5.2  Experimental

The sample was prepared in a solid-state reaction method which has been published elsewhere [11]. In order to perform the $^{17}O$ NMR study, the sample was annealed in the oxygen gas containing 40% of $^{17}O$ at 450°C for 12 h.

The sample used in this study has a nominal composition of $Tl_{1.7}Ba_2Ca_{2.3}$-$Cu_3O_{10}$. $T_c$ of the as-grown sample is 115 K, while it increases to 125 K after annealing in vacuum. In this paper, we report the results on the as-grown sample, while those for the annealed one will be reported in a following paper.

The powder sample was aligned along the $c$-axis in a magnetic field of 11 T and fixed by epoxy for $^{63}Cu$ and $^{17}O$ NMR, while Cu NQR and $^{205}Tl$ NMR have been performed by using a random powder sample.

## 5.3  Electric Field Gradients and Local Hole Distribution

Fig. 5.2(a) shows the $^{63}Cu$ NMR spectra obtained at a field about 11 T for the $c$-axis oriented powder. Two resonance lines have been observed in the direction of $H \perp c$, while they are not resolved in the direction of $H \| c$.

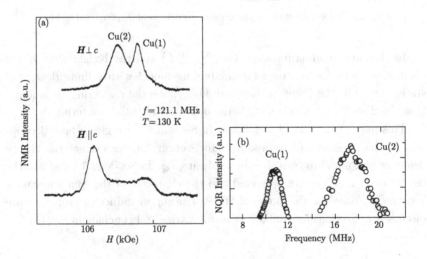

Fig. 5.2  (a) $^{63}Cu$ NMR spectra of partial aligned powder of $Tl_2Ba_2Ca_2Cu_3O_{10}$. (b) Cu NQR spectra obtained at $T = 1.3$ K.

The shift is composed of the Knight shift $K$ and the shift due to quadrupole interaction,

$$\frac{\Delta\nu}{\gamma_N H_{\text{res}}} = K + \frac{3\nu_Q^2}{16(1+K)(\gamma_N H_{\text{res}})^2}. \tag{5.1}$$

In the same way as Takigawa et al. [15], we obtained $\nu_Q$ and the Knight shift for each site. Judging from the fact that the intensity of the left resonance line with $\nu_Q \simeq 17$ MHz is about twice of that of the right line with $\nu_Q \simeq 11$ MHz, we assign the former to arise from the pyramidal $CuO_2$ plane (Cu(2)) and the latter to the square plane (Cu(1)), because their crystal occupancy rate is 2 to 1. The NQR spectra for the two distinct Cu sites, having peaks at 17.4 MHz and 10.8 MHz, respectively, have also been observed, as shown in Fig. 5.2(b). The $\nu_Q$, in agreement with the value already reported [16] is the smallest among the reported materials so far.

Fig. 5.3 shows the $^{17}O$ NMR spectra for $\boldsymbol{H}\|c$-axis and $\boldsymbol{H} \perp c$-axis. Among four sets of the signals (A-D), two (A and B) having the largest component

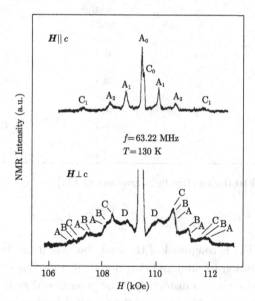

Fig. 5.3   $^{17}O$ NMR spectra for $\boldsymbol{H}\|c$ and $\boldsymbol{H} \perp c$. The peaks (singularities) labeled as A~D correspond to oxygens in pyramidal $CuO_2$, square $CuO_2$, BaO and TlO planes, respectively.

of $\nu_Q$ in the $ab$ direction and shorter $T_1$ are assigned to come from O(1) and O(2). We further assign the set A with larger Knight shift to the oxygen in the pyramidal CuO$_2$ plane, O(2), because Cu(2) has a larger Knight shift (larger spin susceptibility (see Section 5.4)), while we set B to O(1). In the $H\|c$ direction, O(1) and O(4) of the TlO plane are not resolved presumably due to the fact that their $\nu_c$ are close to $\nu_c$ ($2\nu_c$) of O(2). The $\nu_Q$ parameter is shown in Table 5.1.

Table 5.1  Tensors of $\nu$ and the Knight shift for $^{17}$O and $^{63}$Cu at $T = 140$ K.

|  |  | $\nu_\alpha$ (MHz) | $K_\alpha$ (%) |
|---|---|---|---|
| B: O(1) | $\nu_a$ | 1.06 | 0.16±0.01 |
|  | $\nu_b$ | −0.73 | 0.12±0.01 |
|  | $\nu_c$ | — | — |
| A: O(2) | $\nu_a$ | 1.12 | 0.23±0.01 |
|  | $\nu_b$ | −0.77 | 0.13±0.01 |
|  | $\nu_c$ | −0.357 | 0.092±0.005 |
| C: O(3) | $\nu_a$ | −0.66 | 0.03±0.01 |
|  | $\nu_b$ | −0.66 | 0.03±0.01 |
|  | $\nu_c$ | 1.329 | 0.03±0.01 |
| D: O(4) | $\nu_a$ | −0.38 | 0.09±0.01 |
|  | $\nu_b$ | −0.30 | 0.08±0.01 |
|  | $\nu_c$ | — | — |
| Cu(1) | $\nu_a$ | −5.4 | 0.37±0.02 |
|  | $\nu_b$ | −5.4 | 0.37±0.02 |
|  | $\nu_c$ | 10.8 | 1.25±0.02 |
| Cu(2) | $\nu_a$ | −8.7 | 0.51±0.02 |
|  | $\nu_b$ | −8.7 | 0.51±0.02 |
|  | $\nu_c$ | 17.4 | 1.25±0.02 |

$\nu$ is proportional to the electric field gradient (EFG) as

$$\nu_\alpha = \frac{3}{2\hbar I(2I-1)} eQ \frac{\mathrm{d}^2 V}{\mathrm{d}\alpha^2} \quad (\alpha = x, y, z). \tag{5.2}$$

In general the EFG is composed of the contributions from the asymmetrical electronic distribution (onsite) and that due to surrounding charges (lattice). The latter may also cause a distortion of the inner closed shells, which is phenomenologically expressed in terms of the so-called Sternheimer antishielding factor $\gamma_\infty$,

$$eq \equiv \frac{\mathrm{d}^2 V}{\mathrm{d}\alpha^2} = eq_{\text{on-site}} + (1 - \gamma_\infty)eq_{\text{lattice}}. \tag{5.3}$$

$\nu_{\text{lattice}}$ in high-$T_c$ oxides is calculated using a point charge model to be several percent of the observed value. Experimentally, in contrast to the increase of $\nu_Q$ (Cu) on hole doping in La$_{2-x}$Sr$_x$CuO$_4$ [17], 15% doping of electrons in Nd$_{2-x}$Ce$_x$CuO$_4$ wipes out $\nu_Q$(Cu) completely [18]. This suggests that $\nu_Q$ is dominated by the on-site contribution.

Recent theories have demonstrated that it becomes possible to elucidate the origin of $\nu_Q$ from a first principle calculation [19–21]. Schwarz *et al.* [19] and Yu *et al.* [20] have found, from a band calculation using the full potential augmented-plane-wave method, that the lattice effect in the Cu-oxides is negligibly small; $^{63}\nu_Q$ originates from the Cu3$d$ hole and Cu4$p$ electrons, while $^{17}\nu_Q$ comes from O2$p$ holes.

As before [22], we analyze $\nu_Q$ to obtain the hole density in the Cu and oxygen sites in the light of the consequences of the band calculation. The holes at the $\sigma$ orbit (Cu-O bonding) and $\pi$ orbit (others) are [22,23]

$$n_{p\sigma} = (1 + \eta/3)\frac{\nu_a}{|\nu_{2p,0}|}, \tag{5.4}$$

$$n_{p\pi} = \frac{2}{3}\eta\frac{\nu_a}{|\nu_{2p,0}|}, \tag{5.5}$$

where $\eta$ is the asymmetry parameter of $\nu$ and $|\nu_{2p,0}| = 3.646$ MHz is a *one hole* contribution. In this way, $n_{p\sigma}$ in the pyramidal plane is found to be 0.34, which is larger than 0.29 for YBa$_2$Cu$_3$O$_7$.

For the Cu site, the observed $\nu_Q$ is composed of the contribution of 3$d$ holes and 4$p$ electrons [19,20],

$$\nu_Q = n_{x^2-y^2}|\nu_{3d,0}| - |\nu_{4p}|, \tag{5.6}$$

where $\nu_{3d,0}$ is the contribution from one hole in the $d$ orbit, $|\nu_{3d,0}| = 117$ MHz. We apply $|\nu_{4p}| = 69$ MHz corresponding to 0.1 electrons ($e_{4p}$) in both 4$p_x$ and 4$p_y$ orbits which was obtained for YBa$_2$Cu$_3$O$_6$ [24]. Then $n_{x^2-y^2}$ in the pyramidal plane is calculated to be 0.74, which is quite smaller than that in the La system and YBa$_2$Cu$_3$O$_7$ of $\geqslant 0.85$ [22]. This indicates that the covalency between Cu3$d$ and O2$p$ orbits is stronger and the effective Coulomb interaction is weaker in Tl2223.

As seen in Fig. 5.4, the smaller $n_{x^2-y^2}$ is consistent with the smaller orbital Knight shift $K_{\text{orb}}^c$, as expected from the relation

$$K_{\text{orb}} = 2\gamma_e\gamma_n\hbar^2\langle r^{-3}\rangle_{3d} \cdot n_{x^2-y^2} \times 2\mu_B^2 \sum \frac{\langle e|L|x^2 - y^2\rangle^2}{E_e - E_{x^2-y^2}}, \tag{5.7}$$

where $e$ and $L$ denote the excitation state relative to $3d_{x^2-y^2}$ and the angular momentum operator, respectively.

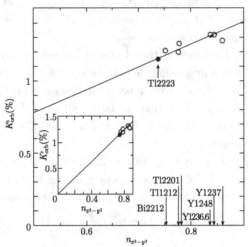

Fig. 5.4   Relation between $K_{\rm orb}^c$ and Cu hole number $n_{x^2-y^2}$. The solid line extrapolates to the origin (also see the inset). The symbols are short for the following materials (see references cited in Ref. [22]). Y1236.6: $YBa_2Cu_3O_{6.6}$, Y1248: $YBa_2Cu_4O_8$, Y1237: $YBa_2Cu_3O_7$, Bi2212: $Bi_2Sr_2CaCu_2O_8$, Tl2201: $Tl_2Ba_2CuO_6$.

It is important to know how many holes, $\delta$, are doped in the $Cu3d_{x^2-y^2}O2p_\sigma$ orbit. In order to estimate $\delta$, we need knowledge about the holes in the $Op_\pi$ orbit. Unfortunately, at present we do not know whether it is due to doping or an effect before doping. Nevertheless, since $n_{p\pi}$ does not depend on $x(= 0.075 - 0.25)$ in $La_{2-x}Sr_xCuO_4$, we tentatively ascribe $n_{p\pi}$ to the consequence before doping. Then $\delta = n_{x^2-y^2} + 2(n)_{p\sigma} - (n_0 - 2(n)_{p\pi})$, where $n_0 = 1 + e_{4px} + e_{4py} = 1.2$ corresponds to the half-filled state with respect to $3d^{10}$ state. So, $\delta$ is 0.378 which is larger than 0.316 for Y1237. As seen for $La_{2-x}Sr_xCuO_4$, $\delta$ may be overestimated by 0.05 by this method.

Fig. 5.5 shows $T_c$ as functions of $\delta$ and the distribution of holes between Cu and oxygen. The hole densities for other compounds than Tl2223 have been calculated in Ref. [22] from the reported $\nu_Q$. Inspection of Fig. 5.5 shows, as we have suggested previously [8,9,22], that one needs not only a proper doping rate but also a strong covalency between Cu and O to obtain a high-$T_c$.

Recently, Hotta [25] and Tsuji *et al.* [5] have discussed the charge distribution from theoretical view points. The relevance of hole distribution with the spin fluctuation spectrum will be mentioned in Section 5.5.

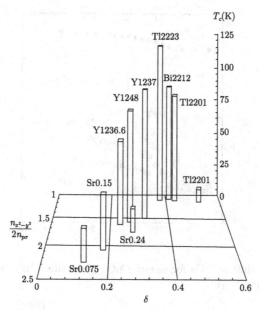

Fig. 5.5  $T_c$ as functions of doping rate $\delta = n_{x^2-y^2} + 2(n_{p\sigma} + n_{p\sigma}) - 1.2$ where 1.2 corresponds to the half-filled state, and the hole distribution between Cu and oxygen, $n_{x^2-y^2}/2(n)_{p\sigma}$, for various materials including Tl2223. Srx% stands for $La_{2-x}Sr_xCuO_4$.

The doping level for the square $CuO_2$ plane is lower than the pyramidal plane, with $n_{x^2-y^2} = 0.68$ and $n_{p\sigma} = 0.33$. However, $n_{x^2-y^2}/2(n)_{p\sigma}$ is quite similar to that for pyramidal plane.

## 5.4   Knight Shifts and Hyperfine Coupling Constants

The $^{63}$Cu Knight shifts for $H$ parallel and perpendicular to the $c$-axis obtained by using Eq. (5.1) are shown in Fig. 5.6. In Fig. 5.7 are shown $^{17}K$ for the O(2) in the pyramidal plane. Our $^{17}K_c$ coincides both in magnitude and $T$

dependence with a shift that Howes *et al.* claimed to be the isotropic part in a random powder sample with similar $T_c$ [26]. The temperature dependence of $^{63}K$ is in agreement with that reported by Han *et al.* [16] where the Knight

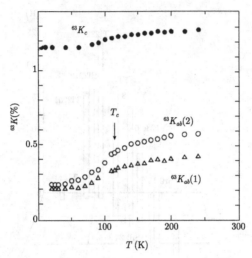

Fig. 5.6   Knight shift of $^{63}$Cu for square $CuO_2$ plane site $(^{63}K(1))$ and pyramidal $CuO_2$ plane site $(^{63}K(2))$.

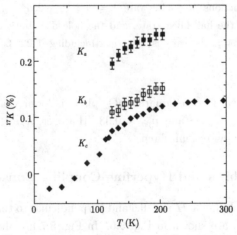

Fig. 5.7   $^{17}$O Knight shifts for O(2) in the pyramidal $CuO_2$ plane.

shift and the quadrupole shift were not separated from each other, but disagrees with Mikhalev $et$ $al.$ [27] who found an abrupt decrease below 130 K but remaining constant above 130 K.

In general the Knight shift $K$ is composed of the spin part $K_s$ and the orbital part, $K_{orb}$, which is temperature independent in general,

$$K = K_s + K_{orb}. \tag{5.8}$$

As seen in Fig. 5.6, $^{63}K_c$ in the normal state decreases with decreasing temperature, indicating that the spin part is finite in contrast to YBa$_2$Cu$_3$O$_7$ where $K_s$ is zero [15]. According to the Mila-Rice Hamiltonian [28] the Knight shifts can be expressed as

$$^{63}K_{s,\alpha} = (A_\alpha + 4B)\chi_0, \quad \alpha = ab, c, \tag{5.9}$$

$$^{17}K_{iso} = 2C\chi_0, \tag{5.10}$$

where $A$ and $B$ are the on-site and the super-transferred hyperfine coupling constants of Cu, respectively, $C$ the transferred hyperfine coupling constants at the O site and $\chi_0$ is the uniform susceptibility. In YBa$_2$Cu$_3$O$_7$, the constants are found to be $B = 40$ kOe/$\mu_B$, $A_c = -160$ kOe/$\mu_B$, $A_{ab} = 37$ kOe/$\mu_B$ and $C = 70$ kOe/$\mu_B$ [28]. $K_{s,c} = 0$ is because $A_{ab} + 4B$ is nearly zero. As seen in Fig. 5.8, $^{63}K_c$ and $^{17}K_{iso} = (^{17}K_a + ^{17}K_b + ^{17}K_c)/3$ have a linear relation with $^{63}K_{ab}$, indicating that they arise from the same spin degree of freedom.

From the fitting in Fig. 5.8, we obtain

$$\frac{A_c + 4B}{A_{ab} + 4B} = 0.47,$$

$$\frac{2C}{A_{ab} + 4B} = 0.50.$$

If we assume that $A_{ab}$ and $A_c$ are the same as in YBa$_2$Cu$_3$O$_7$, we obtain

$$B \simeq 75\text{kOe}/\mu_B,$$

$$C \simeq 80\text{kOe}/\mu_B.$$

$B$ is quite larger than that in YBa$_2$Cu$_3$O$_7$, as is the case in Tl$_2$Ba$_2$CuO$_6$ (Tl2201) [29], and subsequently TlBa$_2$SrCu$_2$O$_7$ (Tl1212) [30] and Bi$_2$Sr$_2$Ca-Cu$_2$O$_8$ (Bi2212) [31], where $B$ is $60 - 100$ kOe/$\mu_B$, while $C$ is comparable to that in $YBa_2Cu_3O_7$ [28]. In the same way, $B$ for Cu(1) is estimated to be similar to Cu(2), thus the difference in magnitude of $^{63}K_{ab}$ between the pyramidal and square planes arises from the difference of $\chi_0$. We wish to note that there is some ambiguity in $^{63}K_c$ due to the overlap of two resonance lines, so our estimation of $B$ should be considered to give the lowest limit. In fact a larger $B$ by 10% would fit the ratio of $^{63}(T_1)_{ab}/^{63}(T_1)_c$ better as seen in the next section. Because $4B$ is much larger than $A_{ab}$, the ratio of $C/B$ which will be used in an analysis of the spin fluctuation character in the next section, however, is little affected by the ambiguity of $^{63}K_c$.

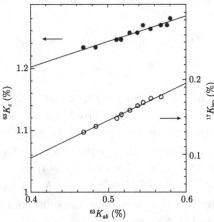

Fig. 5.8   Linear relation between $^{63}K_{ab}$ and $^{63}K_c$ and $^{17}K_{iso}$ for pyramidal $CuO_2$ plane.

The large $B$ seems to be brought about by the strong covalency effect between Cu and O, as found evidence for by the local hole distribution extracted from $\nu_Q$. Since the $B$ term originated from the supertransferred hyperfine interaction from neighbor Cu spins via $Cu3d_{x^2-y^2}$-$O2p_\sigma$-$Cu4(s)$ covalent bond, we may expect a larger hole density in the $Cu4(s)$ orbit for a larger $n_{p\sigma}$.

Finally, we note that the ratio of $^{63}K_c/^{63}K_{ab}$ also puts a constraint on determining $^{63}K_{orb}$. We adopt $^{63}K_c$ at 4.2 K as $^{63}K_{orb,c}(=1.15\%)$, then $^{63}K_{orb,ab}$ is obtained as 0.20%. The finite spin part of $^{63}K_{ab}$ at 4.2 K is due to the pair-breaking effect which will be discussed in Section 5.7. Finally, we note the properties of TlO and BaO planes. As seen in Table 5.1, the Knight shift at the O(4) site (TlO) is quite large, decreasing to 0.03% at $T = 50$ K which is thought to be the orbital contribution. This indicates that the spin susceptibility is finite in the TlO plane, suggesting that this layer is conductive. In contrast, the Knight shift of the O(3) site (BaO) is 0.03% and $T$ independent in the whole $T$ range, indicating that the BaO layer is insulating.

## 5.5  Spin-Lattice Relaxation Rate, Spin–Spin Relaxation Rate and Character of the Spin Fluctuation

$^{63}(1/T_1T)_{ab}$ obtained at the central peaks ($1/2 \longleftrightarrow -1/2$ transition) with the magnetic field perpendicular to the $c$-axis (circles) is shown in Fig. 5.9. At $T = 10$ K, the same value of $T_1$ has been obtained at the peak of the $1/2 \longleftrightarrow 3/2$ transition for both Cu(1) and Cu(2) which do not overlap. We therefore conclude that the quite similar $T_1$ in the two sites are intrinsic, but not due to the effect of partial overlap of each NMR spectrum. For $\boldsymbol{H}\|c$-axis, $T_1$ measured at the central peak of the spectrum has also been shown in the figure (triangle). The anisotropy of the relaxation rate is $R_1 =^{63}(1/T_1)_{ab}/^{63}(1/T_1)_c \simeq 1.8$ for Cu(2) in the normal state, which is quite smaller than 3.7 in YBa$_2$Cu$_3$O$_7$ [32].

$^{17}T_1$ was measured at the peak of the $3/2 \longleftrightarrow 1/2$ transition for O(2) in the direction of $\boldsymbol{H}\|c$. We found a tiny long tail in the magnetization recovery

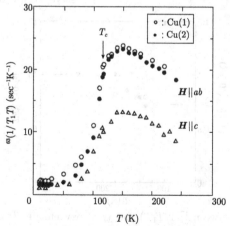

Fig. 5.9  $1/T_1T$ of $^{63}$Cu for $\boldsymbol{H} \perp c$ (circles) and $\boldsymbol{H}\|c$ (triangles).

curve which we ascribe to O(1). $^{17}T_1$ obtained from a good fitting by omitting the slight long component is shown in Fig. 5.10. $T_1$ plotted in logarithmic scale is shown in Fig. 5.11. It is seen that there is a broad peak of $^{63}(1/T_1T)$ above $T_c$. $^{17}(1/T_1T)$ decreases with decreasing $T$, scaling with $\chi_0$ as seen in Fig. 5.12.

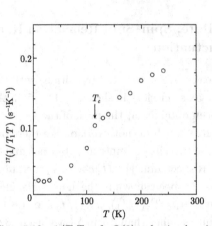

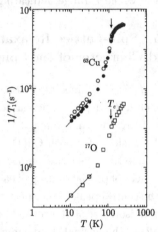

Fig. 5.10  $1/T_1T$ of O(2) obtained with $\boldsymbol{H}\|c$-axis.

Fig. 5.11  $T_1$ at Cu (o: Cu(1) and •: Cu(2)) and O sites plotted in logarithmic scale. The $T_1T$ = const. relation was observed at low temperature.

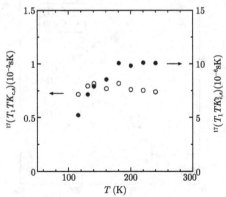

Fig. 5.12  Scaling between $^{17}(1/T_1T)$ and $\chi_0$ at oxygen site. $^{17}(T_1TK_s)$ = const. is more likely. The estimated $K_{c,\mathrm{orb}} = -0.05\%$ was used to obtain $K_{c,s}$.

We have also measured $T_{2G}$, the Gaussian component of the Cu spin-spin relaxation time, which is complementary to $T_1$. As was first shown by Pennington and Slichter [33], $T_{2G}$ in high-$T_c$ cuprates is dominated by the indirect coupling between nuclei via electrons, providing information on the $q$ dependent real part of the electronic susceptibility, $\chi(q)$. In order to obtain a narrower transition line so that most of the spins can be flipped by the $\pi$ pulse, the measurement of $T_{2G}$ was carried out in a field of 5.5 T parallel to the $c$-axis where the half width at full maximum was 150 Oe. The strength of the RF pulse $H_1$ was also about 150 Oe estimated from the width of the RF pulse. The profile of the magnetization decay is shown in Fig. 5.13.

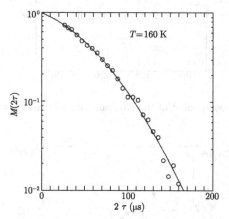

Fig. 5.13   The Cu magnetization decay profile, $t$ is the time interval between the 1st pulse and the echo. The curve is a fit to Eq. (5.11).

The solid curve was a fit to

$$M(t) = M_0 \exp\left[-\frac{1}{2}\left(\frac{t}{T_{2G}}\right)^2 - \frac{t}{T_{2L}}\right], \qquad (5.11)$$

where $t$ is the time interval between the first pulse and echo and $1/T_{2L} = 3(1/T_1)_c + (1/T_1)_{ab}$ [34]; $T_{2G}$ thus obtained is shown in Fig. 5.14. $1/T_{2G}$ obeys a Curie-Weiss relation of

$$1/T_{2G} = \frac{6.6 \times 10^3}{T + 300}(\text{ms}^{-1}).$$

The scaling between $T_1$ and $T_{2G}$ is shown in Fig. 5.15. Above $T = 200$ K, $T_1T/T_{2G}^2$ is nearly constant, while between $T = 150$ K and 200 K, $T_1T/T_{2G}^2 =$ const. seems to be established. This scaling is in a different regime from Y systems [35–37].

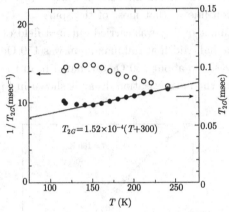

Fig. 5.14   Gaussian component of the Cu spin-spin relaxation rate, $1/T_{2G}$ for pyramidal plane.

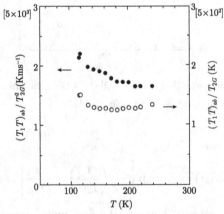

Fig. 5.15   Scaling between $T_1$ and $T_{2G}$. $T_1T/T_{2G}^2 =$ const. is established above 200 K.

Let us discuss the spin dynamics probed by $T_1$ and $T_{2G}$. According to the Mila-Rice Hamiltonian [28], $T_1 = 1/2W$ is written in the forms of

$$^{63}W_c = \frac{3k_B T}{4} \frac{1}{\mu_B^2 \hbar^2} \sum_q F_{ab}^2 \frac{\chi''(q,\omega)}{\omega}, \tag{5.12}$$

$$^{63}W_{ab} = \frac{3k_B T}{8} \frac{1}{\mu_B^2 \hbar^2} \sum_q (F_{ab}^2 + F_c^2) \frac{\chi''(q,\omega)}{\omega}, \tag{5.13}$$

$$^{17}W = \frac{3k_B T}{4} \frac{1}{\mu_B^2 \hbar^2} \sum_q G^2 \frac{\chi''(q,\omega)}{\omega}. \tag{5.14}$$

On the other hand, $T_{2G}$ is expressed as [35,38]

$$\frac{1}{(T_{2G})^2} = \frac{0.69(^{63}\gamma\hbar)^4}{8\hbar^2} \left[ \frac{1}{N} \sum_q F_c^4 \chi(q)^2 - \left( \frac{1}{N} \sum_q F_c^2 \chi(q) \right)^2 \right], \tag{5.15}$$

where $\chi''(\omega, q)$ is the dynamical susceptibility and $\omega$ the NMR frequency. The form factors are as follows:

$$F_{ab} = A_{ab} + 2B(\cos q_x a + \cos q_y a), \tag{5.16}$$

$$F_c = A_c + 2B(\cos q_x a + \cos q_y a), \tag{5.17}$$

$$G = 2C \cos \frac{q_x a}{2}, \tag{5.18}$$

where $a$ is the distance between Cu atoms. In the case of strong antiferromagnetic spin fluctuations,

$$^{63}(1/T_1 T) \sim (A - 4B)^2 \frac{\chi''(Q)}{\omega} \bigg|_{\omega \to 0},$$

$$^{17}(1/T_1 T) \sim 2C^2 \frac{\chi''(0)}{\omega} \bigg|_{\omega \to 0},$$

where $Q$ is the AF wave vector $(\pi/a, \pi/a)$. The ratio of the Cu relaxation rate to the oxygen relaxation rate measures the enhancement of $\chi''(q)$ at $q = Q$ over that at $q = 0$. Fig. 5.16 shows

$$R_2 = \frac{^{63}(1/T_1)_{ab}}{^{17}(1/T_1)} \frac{2(c)^2}{\frac{1}{2}(A_c - 4B)^2 + \frac{1}{2}(A_{ab} - 4B)^2} \frac{^{17}\gamma^2}{^{63}\gamma^2},$$

which is larger than that for $YBa_2Cu_3O_7$ [38]. On the other hand, $^{63}(1/T_1 T)/(A - 4B)^2$ is about half of that for $YBa_2Cu_3O_7$.

In order to extract the detailed features of the spin fluctuations, one needs a physical model for the electronic susceptibility. Moriya, Takahashi and Ueda

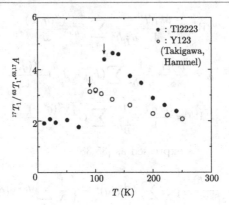

Fig. 5.16  The ratio of the Cu relaxation rate to that of oxygen normalized by the hyperfine coupling constants, which measures the enhancement of spin fluctuation at $q = Q$ over that at $q = 0$. $^{63,17}A = \left\{ 2(c)^2 / \left[ \frac{1}{2}(A_c - 4B)^2 + \frac{1}{2}(A_{ab} - 4B)^2 \right] \right\} \times (^{17}\gamma^2 / ^{63}\gamma^2)$.

(MTU) [2], and Millis, Monien and Pines (MMP) [14] have proposed phenomenological models for $\chi''(q, \omega)$ to extract the parameters of spin fluctuations. We here apply a susceptibility $\chi''(q, \omega)$ of the form of

$$\chi''(q, \omega) = \frac{\pi \chi_0(T) \omega}{\Gamma_0(T)} \times \left[ 1 + \beta(T) \frac{(\xi(T)/a)^4}{[1 + (q - Q)^2 \xi(T)^2]^2} \right], \qquad (5.19)$$

where $\xi$ is the magnetic correlation length, $\Gamma_0$ the characteristic energy of the spin fluctuation at $q = 0$, and $\beta$ measures the strength of the antiferromagnetic (AF) fluctuation $(q \equiv Q = (\pi, \pi))$ relative to the zone-center fluctuations, $\chi_Q = \chi_s \sqrt{\beta} (\xi/a)^2$ and

$$\Gamma_Q = \frac{\Gamma_0}{\sqrt{\beta} \pi (\xi/a)^2}.$$

This model susceptibility is after MMP who treated $\chi_0, \Gamma_0$ and $\beta$ to be temperature independent. Here a Lorentzian form for the energy distribution was assumed, namely,

$$\left. \frac{\chi''(q, \omega)}{\omega} \right|_{\omega \to 0} = \pi \frac{\chi_q}{\Gamma_q}. \qquad (5.20)$$

For such a dynamical susceptibility, the relaxation rates can be expressed as functions of $\chi_0, \Gamma_0, \xi$ and $\beta$,

$$^{63}W_c = \frac{3}{4\mu_B^2\hbar}[(A_{ab} - 4B)^2 S_0 + 8B(A_{ab} - 4B)S_1 + 20B^2 S_2], \qquad (5.21)$$

$$^{63}W_{ab} = \frac{3}{8\mu_B^2\hbar}\{[(A_{ab} - 4B)^2 + (A_c - 4B^2]S_0 + 8B(A_{ab} + A_c - 8B)S_1 + 40B^2 S_2\}, \qquad (5.22)$$

$$^{17}W = \frac{3}{2\mu_B^2\hbar}C^2 S_1, \qquad (5.23)$$

where

$$S_0 = \frac{\pi\chi_0 k_B T}{\hbar\Gamma_0}\left[1 + \beta\left(0.080\left(\frac{\xi}{a}\right)^2 - 0.007\right)\right], \qquad (5.24)$$

$$S_1 = \frac{\pi\chi_0 k_B T}{\hbar\Gamma_0}\left(1 + \beta\left(0.040\ln\frac{\xi}{a} + 0.017\right)\right), \qquad (5.25)$$

$$S_2 = \frac{\pi\chi_0 k_B T}{\hbar\Gamma_0}(1 + 0.026\beta). \qquad (5.26)$$

The spin-spin relaxation rate is given by Thelen and Pines [39] on the basis of Pennington-Slichter theory [33] as

$$\left(\frac{1}{T_{2G}}\right) \simeq \frac{0.69(^{63}\gamma\hbar)^4(A_c - 4B)^4}{32\pi\hbar^2}\beta\chi_0^2\left(\frac{\xi}{a}\right)^2, \qquad (5.27)$$

thus $T_{2G}$ provides information on the magnetic correlation length. For large $\xi$ [39],

$$^{63}\frac{(T_1 T)_c}{T_{2G}^2} \simeq \frac{0.69(^{63}\gamma\hbar)^2 F_c(Q)^2}{16\pi\hbar k_B}(2R_1 - 1)\chi_0\hbar\Gamma_0. \qquad (5.28)$$

From Fig. 5.9, we found $\chi_0\hbar\Gamma_0 = 5.1$ above $T = 200$ K which is larger than 3.5 in $YBa_2Cu_3O_7$ [39]. $\chi_0$ can be extracted from the Knight shift as in Eq. (5.9), which is shown in Fig. 5.17. Then $\Gamma_0$ is obtained as is also shown in Fig. 5.17. $\Gamma_0$ is constant above 200 K, while it is increased below this temperature, mainly due to a deeper decrease of $\chi_0$ below this temperature. With respect to the abrupt change in the characteristic energy of spin fluctuation, we associate with this temperature the onset temperature of a pseudo spin gap which has been intensively argued in the past years (see for example the articles in [40]).

From the $1/T_1$ result, we obtained $\beta/(\xi/a)^2 = 370$ at $T = 115$ K which decreases to 130 at $T = 240$ K. With this constraint, we fit the ratio of Cu relaxation rate to that of oxygen, and the anisotropy of the Cu relaxation rate, $R_1$, to obtain $\xi$ and $\beta$ separately. The result of $\xi$ is plotted in Fig. 5.18.

It increases rapidly with decreasing temperature, while it tends to saturate
around $T_c$. $\beta$ has only a very weak $T$ dependence, varying from 60 to 65. The
former is comparable to but the latter is greater than that in $YBa_2Cu_3O_7$ [39].
With such $\xi$ and $\beta$, $\chi_Q$ and $\Gamma_Q$ are calculated as shown in Fig. 5.19. Thus, the
larger $R_2$ shown in Fig. 5.16 and smaller $^{63}(1/T_1T)/(A - 4B)^2 = \chi_Q/\Gamma_Q\xi^2$
than those in $YBa_2Cu_3O_7$ are explained in terms of larger $\beta$ and $\Gamma_Q$, respectively.

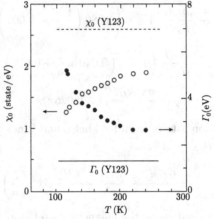

Fig. 5.17   Temperature dependence of $\chi_0$ and the characteristic energy at $q = 0$, $\Gamma_0$.
Also shown are those for $YBa_2Cu_3O_7$ [37].

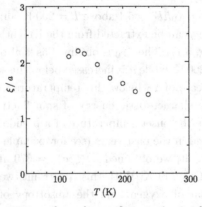

Fig. 5.18   Temperature dependence of magnetic correlation length $\xi$.

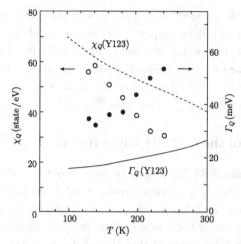

Fig. 5.19   Temperature dependence of the estimated $\chi_Q$ and the characteristic energy of AF spin fluctuations, $\Gamma_Q$. For comparison the results for YBa$_2$Cu$_2$O$_7$ [39] are also shown by the curves.

The larger $\Gamma_0$ and $\Gamma_Q$ indicate that the spin fluctuation spectral weight is transferred to higher energy. Since the AF spin fluctuation is produced by the exchange coupling $J$ between Cu spins, the result means that $J$ is larger in Tl2223. This large $J$ may originate from the weaker Coulomb repulsion $U$ [41] as suggested from the smaller $n_{x^2-y^2}$.

For the mechanism of spin-fluctuation-induced superconductivity, MTU [2] and Monthoux and Pines (MP) [3] showed that the cutoff energy for the effectiveness of attractive interaction scales with the characteristic energy of the AF spin fluctuation, $\Gamma_Q$. In the MP expression [3]

$$T_c = \Gamma_Q \left(\frac{\xi}{a}\right)^2 \frac{1-\delta}{0.79} \exp(-1/\lambda), \qquad (5.29)$$

where $0.42 \leqslant \lambda \leqslant 0.48$. For the present case, $\Gamma_Q \xi^2 = 100$ meV above $T = 200$ K is larger by 40% than 70 meV for YBa$_2$Cu$_3$O$_7$. This accounts for the higher $T_c$ quite well, suggesting that the superconductivity is spin fluctuation mediated. We should also note that the smaller $n_{x^2-y^2}/2n_{p\sigma}$ may favor high-$T_c$ in the mechanism of charge fluctuation-induced superconductivity [5], although the detailed feature of such mechanism has not been clarified yet.

For the role of the square $CuO_2$ plane, we speculate it to act as a trigger to bring about the characteristic properties of the pyramidal plane, since without the square plane, $T_c$ has already reached to 100 K in $Tl_2Ba_2CaCu_2O_8$, while it is not possible so far to obtain a high-$T_c$ without pyramidal plane. The similar $1/T_1T$ for Cu(1) to Cu(2) may arise from the similar $n_{x^2-y^2}/2n_{p\sigma}$, as seen in the relationship between $n_{x^2-y^2}/2n_{p\sigma}$ and $^{63}(1/T_1T)/(A-4B)^2$ [8,22].

## 5.6    Disorder of the Ca/TlO Layer Revealed by $^{205/203}$Tl NMR

Two distinct signals of Tl NMR have been observed in a powder sample as seen in Fig. 5.20. The site (1) shows a positive and $T$-independent $K$, while $K$ of site (2) is negative and dependent on $T$. Crystallographically, Tl occupies a unique site, thus one of them should come from Tl substituted for another atomic site. It has been reported that the off-stoichiometry of Tl and Ca is important for stabilizing the crystal and that Ca and Tl interchanges their crystal position in part [42]. Judging from the results that the Knight shift for site (2) is negative (Fig. 5.21) and scales with $^{63}K_{ab}$ and that the position of Ca resembles that of yttrium in $YBa_2Cu_3O_{6+x}$, we conclude that this signal with negative Knight shift is from Tl that replaced the Ca site, where the supertransferred hyperfine interaction through the Cu3d-O2p-Tl5d bond can

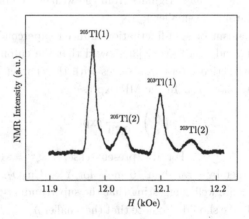

Fig. 5.20   Tl NMR spectra of a random powder sample obtained at $f = 29.5$ MHz and $T = 140$ K.

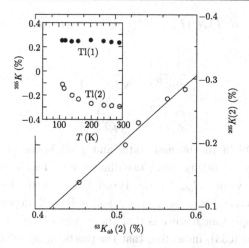

Fig. 5.21   The linear relation between $^{63}K_{ab}$ and $^{205}K(2)$. In the inset is shown the temperature dependence of the Knight shift for two distinct Tl sites.

give rise to a negative $K$. Hentsch *et al.* have reached the same conclusion [43]. This substitution of Ca site by Tl causes a distortion/randomness of the Ca/TlO layer which affects the superconducting properties seriously as seen in the following section.

## 5.7   Superconductivity

In the superconducting state, $1/T_1$ of Cu decreases sharply below $T_c$, as commonly seen in all Cu-oxide superconductors. At lower $T$, however, the sharp decrease of $1/T_1$ is gradually replaced by the $T$ linear dependence below $T \leqslant 25$ K. It should be noted that the $T_1T =$ const. behavior is not due to the presence of vortex cores since no field dependence of $1/T_1$ was observed. In the $T$ range where a $T_1T =$ const. law holds, the Knight shift also becomes constant as seen in Fig. 5.6. Such behaviors, which have previously been reported in Zn substituted $YBa_2Cu_3O_7$ [44], and subsequently in other compounds [45,46,31], are not expected for an $s$-wave superconductor, but are compatible with $d$-wave superconductivity when there is a strong potential scattering which brings about a finite density of states (DOS) at the Fermi surface [47,48]. In the superconducting state, $T_1$ can be expressed as

$$\frac{1}{T_1 T} \propto \int (N_s^2(E) + M_s^2(E))f(E) \times (1 - f(E))dE, \qquad (5.30)$$

where

$$\frac{N_s(E)}{N_0(E)} = \frac{E}{\sqrt{E^2 - \Delta^2}}$$

and

$$\frac{M_s(E)}{N_0(E)} = \frac{\Delta}{\sqrt{E^2 - \Delta^2}}$$

with $N_0$ the DOS in the normal state, and $f(E)$ is the Fermi energy. In Fig. 5.22 is shown a fitting using two-dimensional $d$-wave superconducting DOS with finite value, $N_{\mathrm{res}}$, at the Fermi level, where the superconducting order parameter $\Delta(\theta, \varphi) = \Delta_0 \cos\varphi$ and a BCS $T$ dependence for $\Delta$ is assumed. The important feature is that the residual DOS, $N_{\mathrm{res}}$, is larger for Cu(1) ($N_{\mathrm{res}}/N_0 = 0.33$), indicating that the pairbreaking effects are stronger in this site.

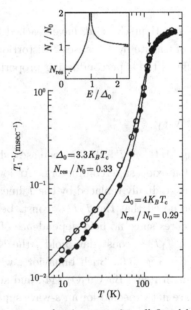

Fig. 5.22  $1/T_1$ in the superconducting state is well fitted by a $d$-wave DOS shown in the inset. The pair breaking effect is larger in the Cu(1) site (○) than in the Cu(2) site (●). The arrow indicates $T_c$.

We propose that the potential scattering in the present case is caused by the disorder of the Ca layer which was described in the previous section. The disorder of the Ca layer itself may scatter the carriers, and/or may also map a distortion onto the $CuO_2$ planes. The larger residual DOS at the square $CuO_2$ plane than at the pyramidal plane supports this identification, since $Cu(1)$ is sandwiched by two Ca layers, while there is only one Ca layer adjacent to $Cu(2)$ (see Fig. 5.1).

The reduction of $T_c$ by the non-magnetic impurity in a $d$-wave superconductor has been investigated theoretically [47,48,3]. According to a recent study by Hotta [46] for the case of cylindrical Fermi surface, a residual DOS of $0.3N_0$ should cause a reduction of $T_c/T_{c0} = 0.87$, where $T_{c0}$ is the transition temperature for a clean sample. This implies that $T_{c0}$ would go up to 132 K if the sample of Tl2223 were perfect in crystal structure. It is interesting that such a $T_{c0}$ is quite close to the record $T_c$ of 135 K realized in $HgBa_2Ca_2Cu_3O_8$ [12] which has a similar crystal structure as Tl2223.

## 5.8   Summary and Conclusion

We have presented a systematic NMR study of the $^{63}Cu$, $^{17}O$ and $^{205}Tl$ in the high-$T_c$ superconductor $Tl_2Ba_2Ca_2Cu_3O_{10}$. By analyzing the nuclear quadrupole frequency $\nu_Q$ at each site, we have obtained the hole numbers at Cu and oxygen in the $CuO_2$ planes. The striking feature is that the covalency between Cu and oxygen is much stronger, i.e. Cu holes are transferred into O site, suggesting a weaker Coulomb repulsion. From the point of view of charge distribution, it is suggested that not only a proper doping rate but also a strong hybridization between Cu and oxygen are favorable for a high-$T_c$. By analyzing the $^{63,17}T_1$ and $^{63}T_{2G}$ using the model of Millis, Monien and Pines, it was found that the spectral weight of the spin fluctuation is transferred to high energy compared to $YBa_2Cu_3O_7$, while the correlation length $\xi$ does not differ much. The larger characteristic energy of spin fluctuations well accounts for its higher $T_c$ in the mechanism proposed by Moriya $et$ $al.$ and Pines $et$ $al.$, suggesting that the superconductivity is spin fluctuation mediated. The superconducting properties are compatible with a $d$-wave superconducting model, with a finite DOS at the Fermi level. It has been found evidence for by Tl NMR that the distortion of Ca/TlO layer caused by partial Tl/Ca inter-substituted acts as the potential scattering. Comparison with the theory by Hotta shows that if such scattering were absent, $T_c$ would go up to 132 K which is quite close to the record $T_c$ realized in $HgBa_2Ca_2Cu_3O_8$.

## Acknowledgments

We thank D. Pines, T. Moriya and K. Ueda for useful discussion, and T. Kuse for assistance in the early stage of this study. This work was partially supported by a Grant-in-Aid for Scientific Research from the Ministry of Education, Science and Culture. The work at ISTEC was partly supported by NEDO through their R&D Industrial Science and Technology Frontier Program.

[1]  *Physical Properties of High-Temperature Superconductivity*, edited by D. M. Ginsberg (World Scientific, Singapore, 1989, 1990).

[2]  T. Moriya, Y. Takahashi and K. Ueda, J. Phys. Soc. Jpn. **59**, 290 (1990); T. Moriya and K. Ueda, J. Phys. Soc. Jpn. **63**, 1871 (1994).

[3]  D. Monthoux and D. Pines, Phys. Rev. B **49**, 4261 (1994).

[4]  J. R. Schrieffer, J. Low Temp. Phys. **99**, 397 (1994).

[5]  T. Tsuji, O. Narikiyo and K. Miyake, Physica C **244**, 311 (1995).

[6]  M. Azami, A. Kobayashi, T. Matsuura and Y. Kuroda, J. Phys. Soc. Jpn. **64** (1995).

[7]  G. Q. Zheng, E. Yanase, K. Ishida, Y. Kitaoka, K. Asayama, R. Tanaka, S. Nakamichi and S. Endo, Solid State Commun. **79**, 51 (1991).

[8]  G. Q. Zheng, T. Kuse, Y. Kitaoka, K. Ishida, S. Ohsugi, K. Asayama and Y. Yamada, Physica C **208**, 339 (1993).

[9]  G. Q. Zheng, T. Mito, Y. Kitaoka, K. Asayama and Y. Kodama, Physica C **243**, 337 (1995).

[10]  Z. Z. Sheng and A.M. Hermann, Nature **332**, 138 (1988).

[11]  T. Kaneko, H. Yamauchi and S. Tanaka, Physica C **178**, 377 (1991).

[12]  A. Schilling, M. Cantoni, J. D. Guo and H. R. Ott, Nature **363**, 56 (1993).

[13]  G. Q. Zheng, Y. Kitaoka, K. Asayama, K. Hamada, H. Yamauchi and S. Tanaka, J. Phys. Soc. Jpn. **64**, 3184 (1995).

[14]  A. J. Millis, H. Monien and D. Pines, Phys. Rev. B **42**, 167 (1990).

[15]  M. Takigawa, P. C. Hammel, R.H. Heffner, Z. Fisk, J.L. Smith and R.B. Schwarz, Phys. Rev. B **39**, 300 (1989).

[16]  Z.P. Han, R. Dupree, R. S. Liu and P. P. Edwards, Physica C **226**, 106 (1994).

[17] See e.g. S. Ohsugi, Y. Kitaoka, K. Ishida and K. Asayama, J. Phys. Soc. Jpn. 60, 2351 (1991).

[18] See e.g. G. Q. Zheng, Y. Kitaoka, Y. Oda and K. Asayama, J. Phys. Soc. Jpn. 58, 1910 (1989).

[19] C. Ambrosch-Draxl, P. Blach and K. Schwarz, J. Phys. Cond. Matt. 1, 4491 (1989); K. Schwarz, C. Ambrosch-Draxl and P. Blach, Phys. Rev. B 42, 2051 (1990); C. Ambrosch-Draxl, E. Blach and K. Schwarz, Phys. Rev. B 44, 5141 (1991).

[20] J. Yu, A.J. Freeman, R. Podloucky, P. Herzig and P. Weiberger, Phys. Rev. B 43, 3516 (1991).

[21] S. Srinivas, S. Sulalman, N. Sahoo, T. E. Das, E. Torikai and K. Nagamine, in Proc. 13th Int. Symp. on NQI (August, 1995, Providence), Z. Naturf., to be published.

[22] G. Q. Zheng, Y. Kitaoka, K. Ishida and K. Asayama, J. Phys. Soc. Jpn. 64, 2524 (1995).

[23] K. Hanzawa, E Komatsu and K. Yosida, J. Phys. Soc. Jpn. 59, 3345 (1990).

[24] Y. Ohta, W. Koshibae and S. Maekawa, J. Phys. Soc. Jpn. 61, 2198 (1992).

[25] T. Hotta, J. Phys. Soc. Jpn. 63, 4126 (1994).

[26] A.P. Howes, R. Dupree, Z.P. Han, R.S. Liu and P.P. Edwards, Phys. Rev. B 47, 11529 (1993).

[27] K. Mikhalev, Yu. Piskunov, Yu. Zhdanov, S. Verkhovskii, A. Trokiner, A. Inyushkin, A. Taldenkov, L. Shustov and A. Yakubovskii, in Proc. 27th Congr. AMPERE (1994, Kazan).

[28] F. Mila and T. M. Rice, Physica C 157, 561 (1989).

[29] Y. Kitaoka, K. Fujiwara, K. Ishida, K. Asayama, Y. Shimakawa, T. Manako and Y. Kubo, Physica C 179, 107 (1991).

[30] G. Q. Zheng, K. Magishi, Y. Kitaoka, K. Asayama, T. Kondo, T. Manako, Y. Shimakawa and Y. Kubo, Physica B 186-188, 1012 (1992); K. Magishi et al., to be submitted.

[31] K. Ishida, Y. Kitaoka, K. Asayama, K. Kadowaki and T. Mochiku, J. Phys. Soc. Jpn. 63, 1104 (1994).

[32] R. E. Waisted, W. W. Warren Jr., R. E. Bell and G.P. Espinosa, Phys. Rev. B 40, 2572 (1989).

[33] C. H. Pennington and C. P. Slichter, Phys. Rev. Lett. 66, 381 (1991).

[34] R. E. Walstedt, Phys. Rev. Lett. 19, 146 (1967).

[35] M. Takigawa, Phys. Rev. B 49, 4158 (1994).

[36] Y. Ito, H. Yasuoka, Y. Fujiwara, Y. Ueda, T. Machi, I. Tomeno, T. Tai, K. Koshizuka and S. Tanaka, J. Phys. Soc. Jpn. 61, 1287 (1992).

[37]  T. Imai, C. P. Slichter, A. P. Paulikas and B. Veal, Phys. Rev. B **47**, 9158 (1993).

[38]  D. Thelen and D. Pines, Phys. Rev. B **49**, 3528 (1994).

[39]  P. C. Hammel, M. Takigawa, R. H. Heffner, Z. Fisk and K. C. Ott, Phys. Rev. Lett. **63**, 1992 (1989); Takigawa *et al.*, Phys. Rev. B **43**, 247 (1991).

[40]  See e.g. *Proc. 4th Int. Conf. on Mater. and Mechanism of Supercond.*, edited by E. Wyder, Physica C **235-240** (1994).

[41]  J. E. Hirsch, Phys. Rev. Lett. **54**, 1317 (1985).

[42]  C. C. Torardi, M. A. Subramanian, J. C. Calabrese, J. Gopalakrishnan, K. J. Morrissey, T. R. Askew, R. B. Flippen, U. Chowdhray and A. W. Sleight, Science **240**, 631 (1988).

[43]  F. Hentsch, N. Winzek, M. Mehring, Hj. Mattausch, A. Simon and R. Kremer, Physica C **165**, 485 (1990).

[44]  K. Ishida, Y. Kitaoka, T. Yoshitomi, N. Ogata, T. Kamino and K. Asayama, Physica C **179**, 29 (1991).

[45]  S. Ohsugi, Y. Kitaoka, K. Ishida, G. Q. Zheng and K. Asayama, J. Phys. Soc. Jpn. **63**, 700 (1994).

[46]  M. Takigawa and Mizi, Phys. Rev. Lett. **73**, 1288 (1994).

[47]  S. Schmitt-Rink, K. Miyake and C. M. Verma, unpublished; K. Miyake, private communication.

[48]  T. Hotta, unpublished.

# 6

# Spectroscopic Evidence for a Pseudogap in the Normal State of Underdoped High-$T_c$ Superconductors*

H. Ding[1,2], T. Yokoya[3], J. C. Campuzano[1,2], T. Takahashi[3], M. Randeria[4], M. R. Norman[2], T. Mochiku[5,6], K. Kadowaki[5,6] and J. Giapintzakis[7]

[1] *Department of Physics, University of Illinois at Chicago, Chicago, Illinois 60607, USA*

[2] *Materials Science Division, Argonne National Laboratory, Argonne, Illinois 60439, USA*

[3] *Department of Physics, Tohoku University, Sendai 980, Japan*

[4] *Tata Institute of Fundamental Research, Bombay 400005, India*

[5] *Institute of Materials Science, University of Tsukuba, Tsukuba, Ibaraki 305, Japan*

[6] *National Research Institute for Metals, Tsukuba, Ibaraki 305, Japan*

[7] *Department of Physics, University of Illinois at Urbana-Champaign, Urbana, Illinois 61801, USA*

It is well known that BCS mean-field theory is remarkably successful in describing conventional superconductors. A central concept of BCS theory is the energy gap in the electronic excitation spectrum below the superconducting transition temperature, $T_c$. The gap also serves as the order parameter: quite generally, long-range phase coherence and a non-zero gap go hand-in-hand [1]. But in underdoped high-$T_c$ superconductors there is considerable evidence that a pseudogap (a suppression of spectral weight) is already formed in the normal state above $T_c$—first, from studies of the spin excitation spectrum [2–5,24], which measure a 'spin gap', and later from a variety of other probes [6–10]. Here we present a study of underdoped $Bi_2Sr_2CaCu_2O_{8+\delta}$

---

* Reprinted from H. Ding, T. Yokoya, J. C. Campuzano *et al.*, Nature **382**, 51 (1996).

(Bi2212) using angle-resolved photoemission spectroscopy (ARPES), which directly measures the momentum-resolved electron excitation spectrum of the $CuO_2$ planes. We find that a pseudogap with $d$-wave symmetry opens up in the normal state below a temperature $T^* > T_c$, and develops into the $d$-wave superconducting gap once phase coherence is established below $T_c$.

In ARPES, photons incident upon a sample cause electrons to be ejected, whose energies and momenta are measured. From this one obtains the electronic excitation spectrum of the sample. Our experiments were carried out at the Synchrotron Radiation Center, Wisconsin, following procedures described previously [11], with 19- or 22-eV photons at an energy resolution of 22–27 meV (full-width at half-maximum, FWHM) and an angular window of $\pm 1°$. Sample underdoping was achieved by adjusting the oxygen partial pressure during annealing of the float-zone grown crystals. These crystals have sharp X-ray diffraction rocking curves with structural coherence lengths of 1250 Å, similar to the near optimally doped samples studied earlier. This high degree of order apparently also helps in maintaining the structural stability of the sample in an ultra-high vacuum of $< 4 \times 10^{-11}$ torr under constant-temperature cycling, even up to 300 K, without measurable sample degradation. All samples show a very flat surface after cleaving, which is essential for determining the intrinsic momentum ($\boldsymbol{k}$) dependence of the gap.

We have studied several samples ranging from overdoped to underdoped, where optimal doping corresponds to a $T_c$ of 92 K. In this Letter we will focus on a moderately underdoped sample with a $T_c$ of 83 K (transition width 2 K), and a heavily underdoped sample with a $T_c$ of 10 K (width $> 5$ K), and contrast these results with a near-optimal (slightly overdoped) 87 K sample (width 1 K). We will simply label the samples by their onset $T_c$s.

It is important to separate the effects of changes in carrier concentration from those due to disorder. Because the superconducting gap for near-optimal Bi2212 is consistent with a $d$-wave order parameter [12–14], for which disorder leads to $T_c$ reduction, we must ensure that the samples we call underdoped have reduced $T_c$ due to lower doping, and not due to increased disorder. In Fig. 6.1 we compare the spectra for the near-optimal (87 K) and underdoped samples with an irradiated sample, whose $T_c$ has been brought down to 79 K by intentionally disordering an 87 K material by high-energy electron irradiation.

We see that the 83 K underdoped sample has a much broader normal-state spectrum that the irradiated sample with a similar $T_c$ (Fig. 6.1(a)), and the 10 K sample has an even broader spectrum. Given the considerable evidence linking ARPES linewidths with electron-electron interactions [15,16], this suggests that underdoped systems are more strongly correlated.

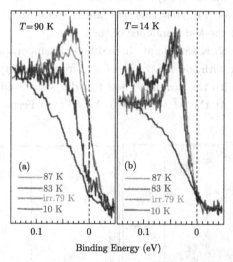

Fig. 6.1   ARPES spectra at the $(\pi, 0) - (\pi, \pi)$ Fermi surface crossing point for four samples (labelled by $T_c$): near-optimal (87 K, red trace), underdoped (83 K and 10 K, bule trace and black trace, respectively), and irradiated (79 K, green trace), at $T = 90$ K (a), and $T = 14$ K (b). Note that all superconducting-state spectra are similar, while the normal-state spectra differ significantly. A comparison of linewidths between the underdoped 83 K and irradiated 79 K samples suggests that disorder is not the main cause of linewidth broadening.

In fact, the spectra of the underdoped sample are so broad that it is difficult to define a spectral peak above $T_c$. It is therefore significant to note in Fig. 6.1(b) that for $T \leqslant T_c$ the line shape sharpens up and a well defined peak is seen for the 83 K sample (at those $k$-points at which there is a sizeable superconducting gap). From Fig. 6.1(b) it is clear that all samples in the superconducting state have very similar low-temperature lineshapes, even though their lineshapes differ significantly in the normal state. We note that the 10 K sample is still in the normal state at $T = 14$ K, and therefore the linewidth sharpening [15] is associated with superconductivity, rather that with the temperature. Because underdoped samples have broad line shapes which are difficult to fit by any simple spectral function [11], we use the shift of the midpoint of the leading edge of a spectrum from the zero of the binding energy (that is, the Fermi energy) as an estimate of the gap. We expect, based on our earlier studies [14], that this estimate is qualitatively correct for both the magnitude and the $k$-dependence of the gap.

Spectra for the 83 K sample at three different temperatures are shown in Fig. 6.2(a) together with spectra (broken lines) from a polycrystalline Pt foil in electrical contact with the sample. The Pt leading-edge midpoint determines the Fermi energy. At the $\overline{M} = (\pi, 0)$ to $Y = (\pi, \pi)$ Fermi surface crossing,

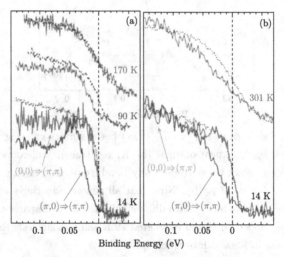

Fig. 6.2   Spectra (solid lines) for underdoped samples at the $(\pi, 0) - (\pi, \pi)$ Fermi surface crossing, along with reference Pt spectra (broken lines) at different temperatures. (a) Spectra (22-eV photons) of the 83 K sample. At 14 K there is a gap in the superconducting state, which persists into the normal state in the 90 K spectrum, eventually closing at $T^* = 170$ K. (b) Spectra (19-eV photons) of the 10 K sample. Note a pseudogap at the $(\pi, 0) - (\pi, \pi)$ Fermi surface crossing exists in the normal state from 14 to 301 K. For both a and b we also plot the spectrum at the $(\pi, 0) - (\pi, \pi)$ Fermi surface crossing at 14 K, showing a vanishing gap.

we find that the leading edge of Bi2212 is pushed to positive binding energies relative to Pt at 14 K indicating a large superconducting gap. This gap does not fully close above $T_c$, as seen in the 90 K data, and we will call this a 'pseudogap'. We use this term because the linewidths above $T_c$ are so broad that there is non-zero spectral weight at the Fermi energy, so that a true gap does not exist. Similar results have also been recently reported in [17] and [18]. The pseudogap persists up to 170 K, which we identify as $T^*$, where no visible shift is seen between the Bi2212 and Pt leading edges and the pseudogap vanishes. We contrast this behaviour to that along the $(\pi, \pi)$ direction, where no gap is seen even well below $T_c$.

Above $T^* = 170$ K, the Fermi surface of the 83 K sample is recovered, by which we mean that we can experimentally map out a closed contour of gapless excitations in $k$-space. Below $T^*$ if one makes a sequence of cuts in $k$-space normal to the Fermi surface, then those points where the gap is smallest along a cut define the minimum gap locus. For the 83 K sample, this locus is found to coincide with the Fermi surface measured above $T^*$, and is a large hole-like barrel enclosing the $(\pi, \pi)$ point. Preliminary results indicate a slight reduction in the enclosed area relative to optimal doping; details will be presented in a future publication.

In Fig. 6.2(b) we show similar results for the 10 K sample. Again, a large pseudogap in the normal state is seen along $\overline{MY}$ and no gap along $\Gamma Y (r = (0, 0))$. For this sample, though, the large pseudogap does not close even at 301 K, the highest temperature at which we did measurements. Thus $T^* > 301$ K at this doping level. As a gapless Fermi surface is not recovered even at 301 K, we use the minimum gap locus (defined above) to infer the shape and size of the Fermi surface which would presumably be recovered above $T^*$. The minimum gap locus for the 10 K sample is a large barrel, as in the optimal and lightly-underdoped cases. Preliminary results indicate that its volume is smaller for the 10 K sample than the 83 K sample, as expected. Note that the existence of a large Fermi surface closed about $(\pi, \pi)$ would contradict several theories of lightly doped Mott insulators which predict small hole pockets centred above $(\pi/2, \pi/2)$.

In Fig. 6.3(a) we plot the angular variation of the leading-edge shifts measured from the Fermi energy at $T = 14$ K along the Fermi surface (FS). For the sake of comparison between samples, we vertically offset the shifts in

Fig. 6.3(a) so that the shift at 45° is zero. (The offset is −3 meV for the 83 K sample and +2 meV for the 10 K sample.) We have seen similar sample-to-sample variations of the offsets for optimal-doped samples with the same $T_c$,

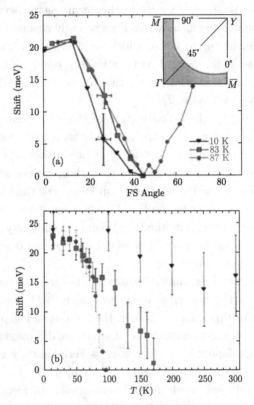

Fig. 6.3   Momentum- and temperature-dependence of the gap estimated from leading-edge shift (see text). (a) $k$-dependence of the gap in the 87 K $T_c$, 83 K $T_c$ and 10 K $T_c$ samples, measured at 14 K. The inset shows the Brillouin zone with a large Fermi surface (FS) closing the $(\pi, \pi)$ point, with the occupied region shaded. Note the striking similarity in both the magnitude and the $k$-dependence of the gap in the three samples with widely differing $T_c$s. (b) Temperature dependence of the maximum gap in a near-optimal 87 K sample (circles), underdoped 83 K (squares) and 10 K (triangles) samples. Note smooth evolution of the 'gap' from superconducting to normal state for the 83 K sample.

so we do not regard these offsets as significant. Note the large error bars on the results for 10 K sample, as this has a rather broad spectrum at 14 K which makes it difficult to accurately determine the midpoint of its leading edge. To address the pseudogap anisotropy for the 10 K sample, we have measured its angular dependence on the locus of minimum gap just as we did for the 83 K sample.

Remarkably, both the magnitude and the $k$-dependence of the gap are similar for the three samples, and independent of carrier concentration, as seen from Fig. 6.3(a), even though $T_c$ and $T^*$ change appreciably. The doping-independence of the pseudogap magnitude has also been suggested in [19]. We note that the observed shift in the 87 K sample corresponds to a super-conducting gap with a maximum value of 33 meV as determined by a spectral function analysis of the data [11,14]. The $k$-dependence of the gap in all the samples is very similar to that of the near-optimal 87 K material, which has been studied in detail by our group [13,14] and found to be consistent with a $d$-wave $| \cos k_x - \cos k_y |$ gap, as suggested in earlier ARPES studies [12].

We emphasize that, as far as ARPES is concerned, there is no difference between the gap seen above and below $T_c$, except in one respect which we discuss below. This is clearly seen for the 83 K sample in Fig. 6.3(b), where we plot the temperature dependence of the leading-edge shifts, without any offsets, at the $M = (\pi, 0)$ to $Y = (\pi, \pi)$ Fermi surface crossing for various samples. (Fig. 6.3(a) and Fig. 6.3(b) were obtained from different data sets, and therefore show slightly different values of the maximum gap). For the 83 K sample there is no signature of the phase transition at $T_c$ in the magnitude of the gap, which eventually vanishes at $T^* = 170$ K. In contrast, in the near-optimal 87 K sample $T^*$ (at which the gap closes) is close to $T_c$. As for the 10 K sample, a pseudogap regime exists over the entire range of temperatures studied. The one difference between the spectra in the superconducting and the pseudogap states is in their linewidths. Above $T_c$ the linewidth is very broad, and spectral weight is present at the Fermi energy even though the leading edge is shifted due to the pseudogap. In contrast, well below $T_c$, the linewidth is resolution-limited, and a clear gap is observed, with very small spectral weight at the Fermi energy due to thermally excited quasi-particles.

These observations are remarkable and, to our knowledge, such effects have not been seen before in ordinary bulk superconductors. Part of our initial motivation was the prediction [20,21] in related model systems that the pseudogap is a normal-state precursor of the superconducting gap above $T_c$, where there is a pairing amplitude without phase coherence [22]. At present, however, there is no quantitative theory to explain our observations of the evolution of the planar Fermi surface, $T^*$, $T_c$ and the gap as functions of the carrier concentration.

We summarize our results in the form of a phase diagram for Bi2212 in Fig. 6.4. At present, this is schematic, as we do not know the precise carrier concentrations for our samples, except for optimal doping, which corresponds approximately to 0.17 holes per Cu [16]. (For hole concentrations in poly-crystalline samples, see [23]) The filled symbols denote the superconducting $T_c$ determined by our susceptibility measurements. The full line (which is a guide to the eye) through these points represents a phase boundary between

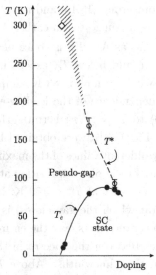

Fig. 6.4 Schematic phase diagram of Bi2212 as a function of doping. The filled symbols are the measured $T_c$s for the superconducting phase transition from magnetic susceptibility. The open symbols are the $T^*$ at which the pseudogap closes; for the $T_c = 10$ K sample, the symbol at 301 K is a lower bound on $T^*$ (thus the hatching). The region between $T^*$ and $T_c$ is an unusual 'normal' state with a pseudogap in the electronic excitation spectrum.

the superconducting (SC) and 'normal' phases. The latter exhibits a pseudogap in its electronic excitation spectrum, which closes only at $T^*$. The experimental points for $T^*$ are denoted by open symbols, and the dashed line through these points denotes a crossover line above which a Fermi surface of gapless excitations is recovered. We find that $T^*$ increases with underdoping, consistent with earlier results [9,10], and 301 K represents only a lower bound on the $T^*$ for our $T_c = 10$ K sample, in which $T_c$ is lower by a factor of at least 30 relative to $T^*$. Given their same $k$-dependence above and below $T_c$, and smooth evolution through $T_c$, our results suggest that the normal-state pseudogap is closely related to the superconducting gap below $T_c$.

This work was supported by the US NSF, the US Dept of Energy Sciences, the US NSF Science and Technology Center for Superconductivity, the Japan Society for the Promotion of Science, NEDO, and the Ministry of Education, Science and Culture of Japan. The Synchrotron Radiation Center is supported by the NSF.

[1]   J. R. Schrieffer, *Theory of Superconductivity* (Cumming, Reading, 1964).

[2]   W. W. Warren *et al.*, Phys. Rev. Lett. **62**, 1193 (1989).

[3]   M. Takigawa *et al.*, Phys. Rev. B **43**, 247 (1991).

[4]   H. Alloul, A. Mahajan, H. Casalta and O. Klein, Phys. Rev. Lett. **70**, 1171 (1993).

[5]   J. Rossat-Mignod *et al.*, Physica B 186-188, 1 (1993).

[6]   C. C. Homes, T. Timusk, R. Liang, D. A. Bonn and W. N. Hardy, Phys. Rev. Lett. **71**, 1645 (1993).

[7]   D. N. Basov, H. A. Mook, B. Dabrowski and T. Timusk, Phys. Rev. B **52**, 13141 (1995).

[8]   J. W. Loram, K. A. Mirza, J. R. Cooper and W. Y. Liang, Phys. Rev. Lett. **71**, 1740 (1993).

[9]   J. L. Tallon, J. R. Cooper, P. de Silva, G. V. M. Williams and J. W. Loram, Phys. Rev. Lett. **75**, 4114 (1995).

[10]   B. Batlogg *et al.*, Physica C **235-240**, 130 (1994).

[11]   H. Ding *et al.*, Phys. Rev. Lett. **74**, 2784 (1995); **75**, 1425 (1995).

[12]   Z. X. Shen *et al.*, Phys. Rev. Lett. **70**, 1553 (1993).

[13]   M. R. Norman, M. Randeria, H. Ding, J. C. Campuzano and A. F. Bellman, Phys. Rev. B **52**, 15107 (1995).

[14]   H. Ding *et al.*, Phys. Rev. B **54**, R9678 (1996).

[15]  M. Randeria *et al.*, Phys. Rev. Lett. **74**, 4951 (1995).

[16]  H. Ding *et al.*, Phys. Rev. Lett. **76**, 1533 (1996).

[17]  A. G. Loeser, Z. X. Shen and D. S. Dessau, Physica C **263**, 208 (1996).

[18]  A. G. Loeser, Science, (in press).

[19]  J. W. Larom, K. A. Mirza, J. M. Wade, J. R. Cooper and W. Y. Liang, Physica C **235-240**, 134 (1994).

[20]  M. Randeria, N. Trivedi, A. Moreo and R. T. Scalettar, Phys. Rev. Lett. **69**, 2001 (1992).

[21]  N. Trivedi and M. Randeria, Phys. Rev. Lett. **75**, 312 (1995).

[22]  V. Emery and S. A. Kivelson, Nature **374**, 434 (1995).

[23]  W. A. Groen, D. M. de Leeuw and L. F. Feiner, Physica C **165**, 55 (1990).

[24]  J. Imai *et al.*, Physica C **162-164**, 169 (1989).

# 7

# Incommensurate Magnetic Fluctuations in YBa$_2$Cu$_3$O$_{6.6}$*

P. C. Dai[1], H. A. Mook[1] and F. Doğan[2]

[1] *Oak Ridge National Laboratory, Oak Ridge, Tennessee 37831-6393*
[2] *Department of Materials Science and Engineering, University of Washington, Seattle, Washington 98195*

We use inelastic neutron scattering to demonstrate that the low-frequency magnetic fluctuations in YBa$_2$Cu$_3$O$_{6.6}$ ($T_c = 62.7$ K) change from commensurate to incommensurate on cooling with the incommensurability first appearing at temperatures above $T_c$. For the energies studied, the susceptibility at incommensurate positions increases on cooling below $T_c$, accompanied by a suppression of the spin fluctuations at the commensurate points. These results suggest that incommensurate spin fluctuations may be a common feature for all cuprate superconductors.

One of the most important questions in the study of high-temperature ($T_c$) cuprate superconductors is the nature of the interplay between the antiferromagnetic (AF) spin fluctuations and superconductivity. Indeed, it is widely believed that the spin dynamical properties of the cuprates are responsible for many of their anomalous transport properties and possibly also the superconductivity. For this reason, the wave-vector ($q$) and energy ($\omega$) dependence of the spin dynamical susceptibility $\chi''(q, \omega)$, which can be probed directly by neutron scattering, has been intensively investigated over the last several years [1–10]. What is puzzling, however, is the variation of the results from one cuprate to another. For the single layer La$_{2-x}$Sr$_x$CuO$_4$ (214) family, spin fluctuations were found at incommensurate positions from the AF lattice point $(\pi, \pi)$ [1,2]. For the bilayer $YBa_2Cu_3O_{7-x}$ [(123)$O_{7-x}$], the situation is less clear. While Rossat-Mignod and co-workers detected only spin fluctuations centered at $(\pi, \pi)$ [3,7], Tranquada $et~al.$ [4] found that the $q$ dependence of the line shape of $\chi''(q, \omega)$ for (123)$O_{6.6}$ is more complex than a simple commensurately centered Gaussian [4]. However, no firm conclusion about the commensurability and symmetry of $\chi''(q, \omega)$ were reached in these experiments. Thus, it is not clear whether the incommensurability in spin fluctuations is specific to the 214 family or an essential property of all cuprate superconductors. A resolution of this issue is important because ultimately, a microscopic theory for high-$T_c$ superconductivity must be able to explain the common features of all cuprate superconductors.

In this Letter, we present inelastic neutron scattering data which resolve the commensurability issue in (123)$O_{6.6}$. We show that the low-frequency spin fluctuations in this material change from commensurate to incommensurate on cooling with the incommensurability first appearing at temperatures above $T_c$. For the energies studied, the susceptibility at incommensurate positions increases on cooling below $T_c$, accompanied by a suppression of the spin fluctuations at the commensurate points. Our results therefore indicate that the incommensurability may be a common feature for all cuprate superconductors.

The neutron scattering measurements were made at the High-Flux Isotope Reactor at Oak Ridge National Laboratory using the HB-1 and HB-3 triple-axis spectrometers. The characteristics of our single-crystal sample of (123)$O_{6.6}$ ($T_c = 62.7$ K) were described in detail previously [8]. The major difficulty in studying spin fluctuations in the (123)$O_{7-x}$ system is to separate the magnetic signal from (single and multi) phonon and other spurious processes. While spurious events such as accidental Bragg scattering can be identified by checking the desired inelastic scan in the two-axis mode [4], two approaches can be used to separate magnetic from phonon scattering. The first approach is to perform neutron polarization analysis [11] which, in principle, allows an unambiguous separation of magnetic and nuclear scattering. This method has been successfully employed to identify the magnetic origin of resonance peaks for ideally [5] and underdoped [8,9] (123)$O_{7-x}$. However, this advantage comes at a considerable cost in intensity which makes the technique impractical for observing small magnetic signals. The second approach is to utilize the differences in the temperature and $q$ dependence of the phonon and magnetic scattering cross sections. While phonon scattering gains intensity on warming due to the thermal population factor, the magnetic signal usually becomes weaker because it spreads throughout the energy and momentum space at high temperatures. Thus, in an unpolarized neutron measurement the net intensity gain above the multiphonon background on cooling at appropriate wave vectors is likely to be magnetic in origin.

Figure 7.1(a) depicts the reciprocal space probed in the experiment with $\mathbf{a}^*(= 2\pi/a)$, $\mathbf{b}^*(= 2\pi/b)$ directions shown in the square lattice notation. The momentum transfers $(q_x, q_y, q_z)$ in units of Å$^{-1}$ are at positions $(H, K, L) = (q_x a/2\pi, q_y b/2\pi, q_z c/2\pi)$ reciprocal lattice units (rlu). We first describe measurements made in the $(H, H, L)$ zone. Our search for the magnetic fluctuations was done with the filter integration technique [12]. This technique is effective for isolating scattering from lower dimensional objects and relies on integrating the energy along the wave-vector direction $[0, 0, L]$ perpendicular to the scan direction $[H, H, 0]$. To estimate the energy integration range of the technique, we note that the scattered intensity for acoustic modulations in (123)O$_{7-x}$ is proportional to the in-plane susceptibility $\chi''(q_x, q_y, \omega)$ [4,13]

$$I(\boldsymbol{q}, \omega) \propto \frac{k_f}{k_i} |f_{\text{Cu}}(\boldsymbol{q})|^2 \sin^2 \left(\frac{1}{2}\Delta z q_z\right) \times [n(\omega) + 1]\chi''(q_x, q_y, \omega),$$

where $k_i$ and $k_f$ are the initial and final neutron wave numbers, $f_{\text{Cu}}(\boldsymbol{q})$ is the Cu$^{2+}$ magnetic form factor, $\Delta z$ $(= 3.342$ Å) the separation of the CuO$_2$ bilayers, $\boldsymbol{q}$ the total momentum transfer $(|\boldsymbol{q}|^2 = q_x^2 + q_y^2 + q_z^2)$, and $[n(\omega) + 1]$ the Bose population factor. The solid line in Fig. 7.1(b) shows the calculated $I(\boldsymbol{q}, \omega)$ at $(\pi, \pi)$ as a function of energy transfer (along $q_z$) assuming $\chi''(q_x, q_y, \omega) = F(q_x, q_y)\chi''(\omega) \propto \omega F(q_x, q_y)$ [14]. Although there are two broad peaks in the figure, the observed intensity will mostly stem from fluctuations around the lower energy one ($10 < \Delta E < 30$ meV) because of the decreased resolution volume at large energy transfers. Since room temperature triple-axis measurements show no detectable magnetic peaks at $(\pi, \pi)$ below $\sim$40 meV (see Figs. 7.2 and 7.3), we have used the integrated scan at 295 K as the background and assumed that the subsequent net intensity gains above the multiphonon background at lower temperatures are magnetic in origin. Figure 7.1(c) shows the result at different temperatures. At 200 K, the magnetic fluctuations are broadly peaked at $(\pi, \pi)$. On cooling to 150 and 100 K, the peak

narrows in width and grows in intensity but is still well described by a single
Gaussian centered at $(\pi, \pi)$. At 65 K, the data show a flattish top similar to
previous observations [4]. Although detailed analysis suggests that the profile
is better described by a pair of peaks (Lorentzian or Lorentzian-squared line
shape) than a single Gaussian, the most drastic change in the profile comes in
the low-temperature superconducting state. Rather than previously observed
single peak, two peaks at positions displaced by $\pm\delta$ (0.057 ± 0.006 rlu) from
$H = 0.5$ are observed, accompanied by a drop in the spin fluctuations at the
commensurate position. The observation of sharp incommensurate peaks with
the filter integration technique suggests that the incommensurability must be
weakly energy dependent in the integration range.

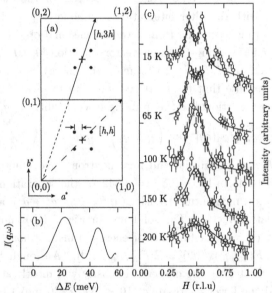

Fig. 7.1   (a) Diagram of reciprocal space probed in the experiment. The dashed
arrow indicates the scan direction with the integrated technique while the solid arrow
represents the triple-axis measurements. (b) Calculated scattered intensity $I(\boldsymbol{q}, \omega)$
as a function of energy transfer. (c) Integrated measurements in which the data at
295 K are subtracted from 200, 150, 100, 65, and 15 K. The data are normalized to
the same monitor count. The solid lines in the 100, 150, and 200 K data are fits to
single Gaussians and linear backgrounds. The solid lines in the 15 and 65 K data are
two Lorentzian-squared peaks on linear backgrounds which best fit the data.

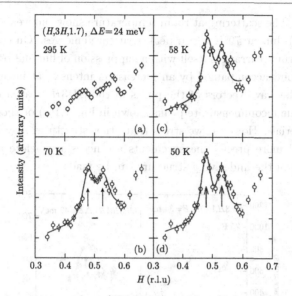

Fig. 7.2   Triple-axis scans along $(H, 3H, 1.7)$ at 24 meV for (a) 295 K, (b) 70 K, (c) 58 K, and (d) 50 K. Data at 295 K were collected with HB-1 while other scans were taken using HB-3. The weak structures in (a) are most likely due to phonon and/or spurious processes. The horizontal bar shows the resolution along the scan direction and the vertical resolution is 0.14 Å$^{-1}$. The positions of incommensurability at $H \approx 0.48$ and 0.53 rlu are indicated by the arrows. Solid lines in (b)–(d) are two Lorentzian-squared peaks on a linear background. The increased scattering at $H > 0.6$ rlu is due to phonons.

Although the integration technique is excellent for finding weak peaks from the scattering of lower dimensional objects, it is important to confirm the result with triple-axis measurements and to determine the symmetry of the incommensurability. For this purpose, we have realigned the sample in the $(H, 3H, L)$ zone. If the 15 K profile in Fig. 7.1(c) stems from an incommensurate structure with peaks at $(0.5 \pm \delta, 0.5 \pm \delta)$ [see Fig. 7.1(a)], scans along the $[H, 3H]$ direction are expected to peak at $H = 0.477$ and 0.523 rlu for $\delta = 0.057$. On the other hand, if the underlying symmetry is identical to that of 214 [rotated 45° from Fig. 7.1(a)], the incommensurability in a $[H, 3H]$ scan should occur at $H = 0.466$ and 0.534 rlu. Figure 7.2 summarizes the result at

24 meV [15]. The scattering at room temperature shows no well-defined peak around $(\pi, \pi)$, but at 70 K a two peak structure emerges. On cooling below $T_c$, the spectrum rearranges itself with a suppression of fluctuations at a commensurate point accompanied by an increase in intensity at incommensurate positions. The wave vectors of the peaks in the $[H, 3H]$ scan are consistent with either the incommensurate peaks shown in Fig. 7.1(a) or those found for the 214 materials. However, we stress that other structures may also explain the data and more precise measurements are necessary before a conclusive identification of the underlying structure can be made.

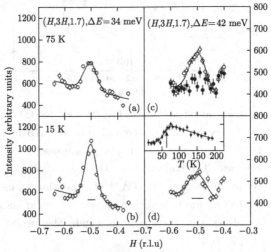

Fig. 7.3   Constant-energy scans along $(H, 3H, 1.7)$ with energy transfer of 34 meV at (a) 75 K, and (b) 15 K. Identical scans at 42 meV at (c) 295 K (•), 75 K (○), and (d) 15 K (○). Inset (•) shows the temperature dependence of the scattering at $(0.5, 1.5, -1.7)$ for $\Delta E = 42$ meV where the arrow indicates $T_c$. The multiphonon background in the 295 K data has been scaled to the value at 75 K for clarity. The horizontal bars represent instrumental resolution. Solid lines are Gaussian fits to the data.

In previous work, superconductivity was found to induce a strong enhancement in $\chi''(\boldsymbol{q}, \omega)$ at $(\pi, \pi)$ for ideally [3,5,6] and underdoped [8–10] (123)$O_{7-x}$ at the resonance positions. Although the intensity gain of the resonance below $T_c$ is accompanied by a suppression of fluctuations at frequencies above it for the underdoped compounds [8,9], no constant-energy scan data are available at energies above the resonance. In light of the present result at 24 meV for the (123)O$_{6.6}$ sample which has a resonance at 34 meV [8], it is important to collect data at these frequencies. Thus, we undertook additional measurements with improved resolution (collimation of 50′-40′-40′-120′) in the hope of resolving possible incommensurability at high energies. Figures 7.3(a) and 7.3(b) suggest that the fluctuations at the resonance energy are commensurate above and below $T_c$ with no appreciable change in width. For an energy above the resonance (42 meV), the scan is featureless at room temperature but shows a well-defined peak centered at $(\pi, \pi)$ at 75 K. Although superconductivity suppresses the magnetic fluctuations [see inset of Fig. 7.3(d)], the $q$ dependence of the line shape cannot be conclusively determined due to the poor instrumental resolution at this energy. Unfortunately, further reduction in resolution volume is impractical due to the concomitant drop in the scattering intensities.

Since the earlier polarized neutron work [8] has shown that for (123)O$_{6.6}$ the 34 meV resonance is the dominant feature of $\chi''(q, \omega)$ at $(\pi, \pi)$ in the low-temperature superconducting state, it is important to compare the newly observed incommensurate peaks to the intensity gain of the resonance. Figure 7.4 shows the difference spectra between 15 and 75 K at frequencies below and above the resonance. In the energy and temperature range of interest (15 to 75 K), the phonon scattering changes negligibly and the Bose population factor $[n(\omega) + 1]$ modifies the scattered intensity at high temperatures by only 3% at 24 meV and less at higher energies. Therefore, the difference spectra in the figure can be simply regarded as changes in the dynamical susceptibility, i.e., $\chi''(15\text{ K}) - \chi''(75\text{ K})$. Inspection of Figs. 7.4(a) and 7.4(b) reveals that the susceptibility at the incommensurate positions increases on cooling from the normal to the superconducting state, accompanied by a suppression of fluctuations at the commensurate point. Comparison of Fig. 7.4(c) to Figs. 7.4(a) and 7.4(b) indicates that the net gain in intensity at the incommensurate positions below $T_c$ is much less than that of the resonance. For an energy transfer of 42 meV, the intensity drop appears uniform throughout the measured profile; however, instrumental resolution may mask any possible incommensurate features. Figure 7.4(e) plots a summary of the triple-axis measurements in superconducting state. Although there are only two constant-energy scans for

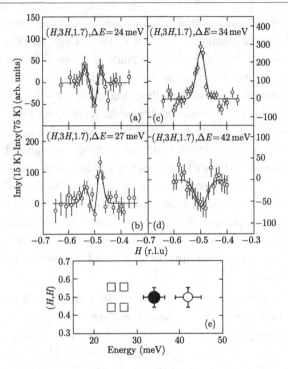

Fig. 7.4   Difference spectra along $(H, 3H, 1.7)$ between low temperature ($< T_c$) and high temperature ($\approx T_c + 12$ K) at (a) 24 meV, (b) 27 meV, (c) 34 meV, and (d) 42 meV. All data were taken with the same monitor units. Solid lines are guides to the eye. (e) Summary of triple-axis measurements. Open squares indicate incommensurate positions. Solid and open circles are the resonance and fluctuations at 42 meV, respectively. The error bars show the energy resolution and the intrinsic $q$ width (FWHM).

frequencies below the resonance, these data nevertheless confirm the result of the integrated technique.

To place our work in proper context, it is useful to compare the results with various theoretical predictions. If the $(123)O_{7-x}$ system is indeed a $d$-wave superconductor, $d$-wave gap nodes could yield incommensurate peaks [16] and the scattering of $(123)O_{6.6}$ is expected to change from commensurate in the normal state to incommensurate in the superconducting state [17]. In this scenario, the susceptibility at incommensurate positions should increase on cooling below $T_c$ for all frequencies below the $d$-wave gap [17,18]. Our preliminary triple-axis measurements [19] show that spin fluctuations at 16 meV are also incommensurate in the normal state, but on cooling below $T_c$ these fluctuations are suppressed. At present, it remains unclear how to reconcile *this* simple $d$-wave picture with these results, or perhaps even more problematical, with the appearance of the incommensurate fluctuations at temperatures above $T_c$.

Alternatively, the observed incommensurate peaks may be viewed as the signature of a stripe phase. Tranquada *et al.* [20] have argued that the incommensurability in 214 may be associated with the spatial segregation of charge or charge density wave correlations [21]. However, if the idea of dynamical microphase separation in the CuO$_2$ plane asserted by Emery and Kivelson [22] is relevant for the high-$T_c$ superconductivity, one would expect incommensurate spin fluctuations in other cuprate superconductors. The observation of such fluctuations in (123)O$_{7-x}$ is consistent with this picture. Unfortunately, there are no explicit predictions about the incommensurate structure in (123)O$_{7-x}$ from a stripe model that can be directly compared with our experiments.

Finally, previous interpretations of NMR experiments have assumed that AF spin fluctuations in (123)O$_{7-x}$ are commensurate and the spin correlation length $\xi$ is temperature independent [23]. Our data suggest a reconsideration of these assumptions, particularly in view of the apparent contradiction between the results of NMR and neutron scattering experiments [24].

We thank G. Aeppli, V. J. Emery, S. M. Hayden, K. Levin, and D. Pines for helpful discussions. We have also benefited from fruitful interactions with J. A. Fernandez-Baca, R. M. Moon, S. E. Nagler, and D. A. Tennant. This research was supported by the U.S. DOE under Contract No. DE-AC05-96OR22464 with Lockheed Martin Energy Research Corp.

[1] S. W. Cheong *et al.*, Phys. Rev. Lett. **67**, 1791 (1991); T. E. Mason *et al., ibid.* **77**, 1604 (1996).

[2] K. Yamada *et al.*, Phys. Rev. Lett. **75**, 1526 (1995).

[3] J. Rossat-Mignod *et al.*, Physica **185C**, 86 (1991).

[4] J. M. Tranquada *et al.*, Phys. Rev. B **46**, 5561 (1992); B. J. Sternlieb *et al., ibid.* **50**, 12 915 (1994).

[5]  H. A. Mook *et al.*, Phys. Rev. Lett. **70**, 3490 (1993).

[6]  H. F. Fong *et al.*, Phys. Rev. Lett. **75**, 316 (1995).

[7]  L. P. Regnault *et al.*, Physica **213B**, 48 (1995).

[8]  P. Dai *et al.*, Phys. Rev. Lett. **77**, 5425 (1996).

[9]  H. F. Fong *et al.*, Phys. Rev. Lett. **78**, 713 (1997).

[10] P. Bourges *et al.*, Europhys. Lett. **313**, 38 (1997).

[11] R. M. Moon *et al.*, Phys. Rev. **181**, 920 (1969).

[12] H. A. Mook *et al.*, Phys. Rev. Lett. **77**, 370 (1996).

[13] The time-of-flight measurements at the ISIS pulsed spallation source [see, for example, S. M. Hayden *et al.*, Phys. Rev. B **54**, R6905 (1996); and arXiv: cond-mat/9710181] on the sample show that the optical modes appear first at $60\pm5$ meV, thus it is sufficient to model the magnetic fluctuations with the acoustic modulation for the energy range probed in the present experiment.

[14] Here we have adopted the notation of Ref. [4]. The expression $\chi''(\omega) \propto \omega$ is valid for $\omega \to 0$ (see, for example, Ref. [24]) and we have extended it to give an estimate of the integration range.

[15] In Ref. [8], we have observed a drop in magnetic signal at $(\pi,\pi)$. Unfortunately, the polarized constant-energy scan at 24 meV [see Fig. 2(b) of Ref. [8]] did not pick up the incommensurate feature described in this work due to limited intensity of the technique.

[16] J. P. Lu, Phys. Rev. Lett. **68**, 125 (1992).

[17] Y. Zha *et al.*, Phys. Rev. B **47**, 9124 (1993).

[18] N. Bulut and D. J. Scalapino, Phys. Rev. B **47**, 3419 (1993).

[19] P. Dai, H. A. Mook and F. Doğan, Physica B (to be published).

[20] J. M. Tranquada *et al.*, Nature **375**, 561 (1995); Phys. Rev. B **54**, 7489 (1996); Phys. Rev. Lett. **78**, 338 (1997).

[21] J. Zaanen and O. Gunnarsson, Phys. Rev. B **40**, 7391 (1989); U. Löw *et al.*, Phys. Rev. Lett. **72**, 1918 (1994); C. Castellani *et al.*, *ibid.* **75**, 4650 (1995).

[22] V. J. Emery and S. A. Kivelson, Physica **209C**, 597 (1993); **235**, 189 (1994).

[23] See, for example, J. Bobroff *et al.*, Phys. Rev. Lett. **79**, 2117 (1997).

[24] Y. Zha *et al.*, Phys. Rev. B **54**, 7561 (1996); C. Berthier *et al.*, J. Phys. I **6**, 2205 (1996).

# 8

# Microscopic Electronic Inhomogeneity in the High-$T_c$ Superconductor Bi$_2$Sr$_2$CaCu$_2$O$_{8+x}$*

S. H. Pan[1], J. P. O'Neal[1], R. L. Badzey[1], C. Chamon[1], H. Ding[2], J. R. Engelbrecht[2], Z. Wang[2], H. Eisaki[3], S. Uchida[3], A. K. Gupta[4], K.-W. Ng[4], E. W. Hudson[5], K. M. Lang[5] and J. C. Davis[5]

[1] *Department of Physics, Boston University, Boston, Massachusetts 02215, USA*

[2] *Department of Physics, Boston College, Chestnut Hill, Massachusetts 02467, USA*

[3] *Department of Superconductivity, University of Tokyo, Yayoi, 2-11-16 Bunkyo-ku, Tokyo 113-8656, Japan*

[4] *Department of Physics and Astronomy, University of Kentucky, Lexington, Kentucky 40506-0055, USA*

[5] *Department of Physics, University of California, Berkeley, California 94720, USA*

The parent compounds of the copper oxide high-transition-temperature (high-$T_c$) superconductors are unusual insulators (so-called Mott insulators). Superconductivity arises when they are 'doped' away from stoichiometry [1]. For the compound Bi$_2$Sr$_2$CaCu$_2$O$_{8+x}$, doping is achieved by adding extra oxygen atoms, which introduce positive charge carriers ('holes') into the CuO$_2$ planes where the superconductivity is believed to originate. Aside from providing the charge carriers, the role of the oxygen dopants is not well understood, nor is it clear how the charge carriers are distributed on the planes. Many models of high-$T_c$ superconductivity accordingly assume that the introduced carriers are distributed uniformly, leading to an electronically homogeneous system as in ordinary metals. Here we report the presence of an electronic inhomogeneity in Bi$_2$Sr$_2$CaCu$_2$O$_{8+x}$, on the basis of observations using scanning tun-

---

* Reprinted from S. H. Pan, J. P. O'Neal, R. L. Badzey *et al.*, Nature **413**, 282 (2001).

nelling microscopy and spectroscopy. The inhomogeneity is manifested as spatial variations in both the local density of states spectrum and the superconducting energy gap. These variations are correlated spatially and vary on the surprisingly short length scale of $\sim$14 Å. Our analysis suggests that this inhomogeneity is a consequence of proximity to a Mott insulator resulting in poor screening of the charge potentials associated with the oxygen ions left in the BiO plane after doping, and is indicative of the local nature of the superconducting state.

We have carried out extensive low-temperature scanning tunnelling microscopy/spectroscopy (STM/S) studies on optimally doped $Bi_2Sr_2CaCu_2O_{8+x}$. The technical details of the experiments have been reported previously [2,3]. Various single crystal samples fabricated with different techniques (directional-flux solidification or floating-zone) have been used to ensure the generality of the observations. Some crystals are nominally pure and some are deliberately doped with a very dilute concentration of impurity atoms (Zn or Ni). We have consistently observed a particular type of electronic inhomogeneity regardless of the differences between the individual crystals. This inhomogeneity is manifested as spatial variations in the local density of states (LDOS) spectrum, in the low-energy spectral weight, and in the magnitude of the superconducting energy gap.

As an example, Fig. 8.1(a) presents a topographic image obtained on a pure $Bi_2Sr_2CaCu_2O_{8+x}$ crystal grown by the flux-solidification method. The image reveals the crystal structure of the BiO plane with atomic resolution accompanied by the well known incommensurate structural modulations. In addition, an inhomogeneous background is also discernible. After removing the contrast associated with the above-mentioned topological structures by Fourier filtering, this inhomogeneous background is clearly visible. Assuming that the tunnelling matrix element has no spatial dependence, the resulting image shown in Fig. 8.1(b) can be approximated by a map displaying the variation of the integrated LDOS. With spatially resolved spectroscopy, we find that the local tunnelling spectrum also varies from location to location. We discover that the magnitude of the superconducting gap extracted from the local tunnelling spectrum varies spatially as well, instead of exhibiting the single-valued nominal gap of $\sim$40 meV that is usually observed by tunnelling in crystals with optimal oxygen doping [4,5]. Both the integrated LDOS and the superconducting gap magnitude vary on an apparently much shorter length scale than those in the phenomena observed in earlier experiments [6,7]. These observations are surprising because in conventional Bardeen–Cooper–Schrieffer (BCS) superconductors (such as Nb) the integrated LDOS, the tunnelling spectrum, and the superconducting gap are spatially homogeneous.

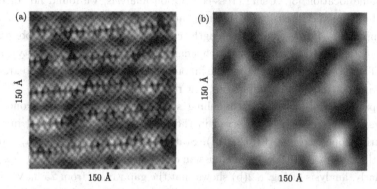

Fig. 8.1   Topographic image and associated integrated LDOS map of an optimally oxygen-doped, nominally pure single crystal of $Bi_2Sr_2CaCu_2O_{8+x}$. (a) Constant-current-mode, topographic image (150 Å×150 Å) of the surface BiO plane exposed after cleavage of the single crystal. In addition to the clear contrast due to the topological corrugations of the atomic structures and the well known incommensurate structural modulation, an inhomogeneous background is also visible. In constant-current mode, the tunnelling current is exponentially related to the tip-sample distance and is also proportionally related to the integrated LDOS. Therefore, the constant current topographic image provides the convolved information of both the topology and the LDOS of the crystal surface. (b) To reveal the inhomogeneous background more clearly, we use Fourier filtering to remove the contrast due to the two well-ordered topological structures mentioned above. The variation of the integrated LDOS, which is seen as an inhomogeneous background in (a) is now clearly displayed. A brighter colour represents a larger magnitude of the integrated LDOS.

What is the origin of such an inhomogeneity? Is it intrinsic or caused by impurities such as crystal defects or excess elements that limit the sample quality? Figure 8.2 is an example of the STM/S results obtained on high quality, floating-zone-grown single crystals deliberately doped with a very dilute concentration (0.2%) of Zn atoms. The inhomogeneous background observed in these crystals (Fig. 8.2(a)) is of the same type as that seen in Fig. 8.1(b). This suggests that the inhomogeneity does not originate from impurities. To support this point, we identified the locations of the Zn impurities from a zero bias conductance map, which is taken simultaneously with the LDOS map at the same location [3]. Using cross-correlation analysis, we found no correlation between the intensity of the integrated LDOS and the locations of the Zn impurities. Furthermore, the length scale of the spatial variation observed in the integrated LDOS map is much smaller than the average impurity spacing. Therefore, we conclude that the inhomogeneity observed in the integrated LDOS is not induced by impurities, but rather is intrinsic in nature.

Spatial variations of the tunnelling spectrum and of the superconducting gap, similar to those observed in the pure sample, are observed in this impurity-doped sample (Fig. 8.2(b)). Statistical analysis of Fig. 8.2(a) shows that the integrated LDOS has a Gaussian distribution, displayed in Fig. 8.2(c). Similarly, analysis of Fig. 8.2(b) shows that the gap ranges from 25 meV to 65 meV and exhibits a Gaussian distribution (42 meV mean; ~20 meV full-width at half-maximum, FWHM) as shown in Fig. 8.2(d). We note that the average gap value is very similar to the single value reported on optimally doped $Bi_2Sr_2CaCu_2O_{8+x}$ in earlier tunnelling measurements [4,5].

In addition to the similarity of their statistical distributions, a strong spatial correlation between the integrated LDOS and the superconducting gap can be seen readily by recognizing similar patterns in the two maps (Fig. 8.2(a) and (b)), which are obtained at the same location simultaneously. As illustrated in Fig. 8.2(e)-(g), auto-correlation analysis on both maps shows similar decay lengths $\xi$ of approximately 14 Å and cross-correlation shows a pronounced Gaussian peak, confirming the local nature of the variations and the strong correlation of these two quantities. Furthermore, the line-shape of the local tunnelling spectrum also correlates with the integrated LDOS. As the gap and the integrated LDOS variations are correlated and the energy gap variation cannot be attributed to a tunnelling matrix effect, the inhomogeneity in the integrated LDOS we observe is most probably intrinsic to the electronic structure and not due to a spatially varying matrix element.

Spatial variation of the energy gap and its correlation with the LDOS can be seen in greater detail in Fig. 8.3. Clearly, the spectra obtained at points

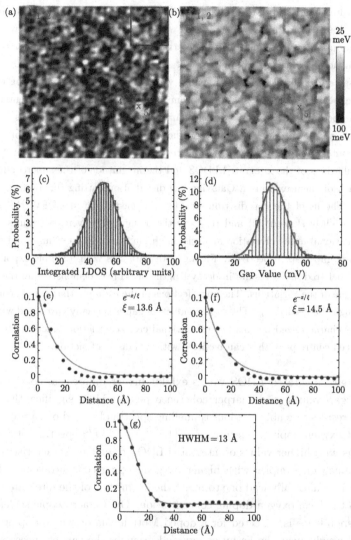

Fig. 8.2   A comparison of an integrated LDOS map and its corresponding superconducting gap map, including their associated statistical results. (a) A 600 Å×600 Å LDOS map obtained with the same technique used in Fig. 8.1(b) on a single crystal $Bi_2Sr_2CaCu_2O_{8+x}$ doped with a very dilute concentration of Zn atoms. (The Zn concentration measured by STM is 0.2% [2]) The crystal has a superconducting transition

temperature of 84 K, with a transition width of 4 K. The red box in panel (a) frames a 150 Å×150 Å region for easy comparison with Fig. 8.1(b). It clearly displays an inhomogeneous structure similar to that observed in Zn-free crystals in Fig. 8.1(b). (b) Superconducting gap map, obtained simultaneously with the integrated LDOS map on the same location, showing the spatial variation of the superconducting energy gap. The local gap values are extracted from the corresponding local differential conductance spectra. We assigned reverse colour coding (higher intensity corresponds to a smaller gap magnitude) to the map so that it is easier to visualize its correlation with the integrated LDOS map. (c) and (d) are the histograms showing the statistical distributions of the integrated LDOS and the magnitude of the superconducting gap. Each of them exhibits a Gaussian-like distribution (fitting function displayed in red). The fit of the gap distribution (42 meV mean; ~20 meV FWHM) shows it to be slightly skewed. (e) and (f) show the azimuthally averaged results of the two-dimensional auto-correlation analysis on the integrated LDOS map and superconducting gap map respectively. Fitting both data sets with a simple exponential decay model (red lines) results in decay lengths of ~14 Å, demonstrating the short length scale of the variations. The imperfections of the fit imply that a more complex model might be needed. (g) The superconducting gap is spatially correlated with the LDOS as characterized by this two-dimensional cross-correlation function. It has a pronounced centre peak that can be fitted with a Gaussian function.

with larger integrated LDOS values exhibit higher differential conductance, smaller gap values, and sharper coherence peaks. More significantly, these STM/S results resemble previous tunnelling spectra obtained on samples with different oxygen doping concentrations [4,5]—the STM/S spectra obtained at positions with higher values of integrated LDOS resemble characteristic spectra obtained on samples with higher oxygen doping concentrations. These observations naturally lead one to relate the magnitude of the integrated local DOS to the local oxygen doping concentration. On a macroscopic scale, such a correspondence is expected for a doped Mott insulator where doping adds spectral weight near the Fermi energy and provides the carriers necessary to transform the insulating compound into a superconductor, as observed previously by photoemission [8]. The idea of local doping concentration (LDC) extends this picture to microscopic scales. In the LDC framework, the charge potentials of oxygen dopants alter the local electronic environment provided

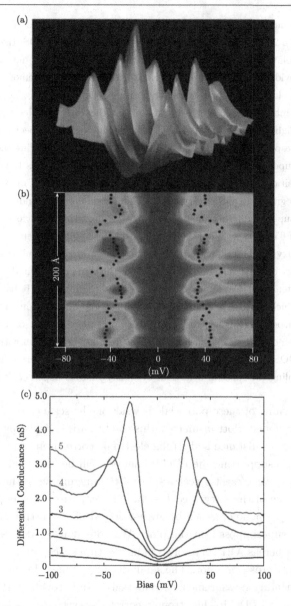

Fig. 8.3  Spatial variation of the tunnelling differential conductance spectrum. To account for variations in the tunnelling junction, the spectra are normalized to a

constant tip-sample separation. (a) A three-dimensional rendering of the tunnelling spectra along a 200Å line, showing detailed variations of the LDOS, the energy gap, and their correlation. We note that the coherence peak heights vary more dramatically than the gap widths. (b) The same data as a bird's-eye view. To demonstrate the gap variation, the black dotted lines trace the positions of the coherence peaks. Together with the gap map of Fig. 8.2(b), we can see that the gap varies less rapidly within a 'patch' of high DOS than at its edge. (c) Five characteristic spectra taken at the positions correspondingly marked in both Fig. 8.2(a) and (b) showing the correlation between the superconducting gap and the integrated LDOS. Curves 1 and 2 are taken at nearby positions in a very dark area on the integrated LDOS map in Fig. 8.2a, where the integrated LDOS is very small. The low differential conductance and the absence of a superconducting gap are indicative of insulating behaviour. Curve 3 has a large gap of 65 meV, with low coherence peaks, resembling the spectral line-shape measured in oxygen-underdoped samples [4,5]. The integrated value of the LDOS at the position for curve 3 is small but larger than those in curves 1 and 2, as it is in a slightly brighter area. Curve 4 comes from a still brighter area. It has a gap value of 40 meV, which is close to the mean value of the gap distribution. The coherence peaks are sharper and higher, resembling the spectral behaviour of a sample with optimal oxygen doping [4,5]. Curve number 5, taken at the position with the highest integrated LDOS, has the smallest gap value of 25 meV with two very sharp coherence peaks, resembling the spectral behaviour of an oxygen over-doped crystal [4,5].

that the screening of these potentials is weak on the scale of the inter-carrier distance. For doped Mott insulators, this condition is satisfied by a relatively large inter-carrier distance and strong electronic correlation. A detailed theoretical analysis supporting the LDC framework will be discussed elsewhere.

In Fig. 8.4, we present a scatter-plot of the magnitude of the energy gap versus the value of the integrated LDOS. For comparison, the doping dependence of the gap value obtained by angle-resolved photoelectron spectroscopy (ARPES) is superimposed on the plot. We note that STM is a real-space, local probe, whereas ARPES resolves in momentum space but averages in real space over a macroscopic spot. The similarity in the linear behaviour of these two complementary measurements is remarkable, and provides further support for the concept of LDC in this strongly correlated system.

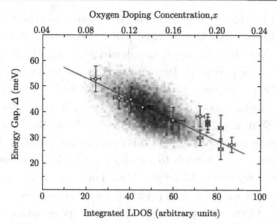

Fig. 8.4   A scatter plot of the superconducting gap versus integrated LDOS. The 16384 data points of the corresponding integrated LDOS map and the gap map (Fig. 8.2(a), (b)) are plotted here in colour. A darker colour represents a larger number of data points with the same integrated LDOS and the same gap magnitude. For comparison, a set of data from ARPES measurements on various $Bi_2Sr_2CaCu_2O_{8+x}$ single crystals with different oxygen doping concentrations is superimposed onto the plot. The gap values of ARPES data are the maximum values of the $d$-wave gap at $(\pi, 0)$ of the Brillouin zone. They are presented as open circles in the plot and the corresponding doping concentration scale is placed on top of the plot. The solid black line is the linear fit to both the STM and the ARPES data.

The existence of such microscopic inhomogeneities should have many important consequences on the quasiparticle properties that are accessible by macroscopic measurements. For example, unexpectedly broad peaks in an earlier neutron scattering measurement on $Bi_2Sr_2CaCu_2O_{8+x}$ could possibly be explained by the presence of inhomogeneity in the bulk [9]. However, it is more important to compare our results to a complementary technique such as ARPES, which measures the same electronic excitations in momentum space under similar experimental conditions. In doing so, we find that taking a spatial average of the STM results yields the same maximum gap value and the same doping dependence of the gap as obtained by ARPES. In addition, the large intrinsic width ($\sim$20 meV) of the coherence peak measured by ARPES [10,11] near the antinodes is consistent with the 20 meV FWHM distribution of the superconducting gap shown in Fig. 8.2(d). In light of the electronic disorder observed by this STM experiment, the discrepancy between the nodal and anti-nodal quasiparticle mean free paths observed by ARPES can now be reconciled. Near the antinodes, gap disorder can significantly scatter the quasiparticles at the gap edge. Indeed, the peak in the momentum distribution curve (MDC) near the antinodes has a large width of $\sim 0.08$ Å$^{-1}$ (ref. 12), corresponding to a short mean free path of $\sim$25 Å. In the nodal direction, however, ARPES measures a mean free path of $\sim$100 Å [13,14]. The quasiparticles of the $d$-wave superconductor in this direction are less affected by the gap disorder since the latter amounts to velocity disorder in the dispersion, which is not effective at scattering Dirac particles. Thus, the mean free path of the nodal quasiparticles will be limited primarily by elastic scattering from the potential disorder. Given that the oxygen dopants responsible for this potential disorder do not reside in the CuO$_2$ plane, scattering is relatively weak. This can result in a longer mean free path and the measured 100 Å length scale is therefore quite reasonable. As scattering into all angles is involved in ARPES measurements, whereas transport is dominated by large-angle scattering of the nodal quasiparticles, our results may also reconcile the discrepancy between the mean free path measured by transport [15] as compared to that measured by ARPES. Taking 14 Å as the length scale over which the disorder potential varies significantly, we deduce that the dominant elastic scattering process is limited by the wavevector $q = 1/14$ Å$^{-1}$. The scattering angle, $\theta$, given by $\sin(\theta/2) = q/2k_{\mathrm{F}}$, where $k_{\mathrm{F}}$ is the Fermi wavevector, is indeed quite small at about 5°, which provides a possible explanation as to why the transport mean free path can be much longer than that measured by ARPES.

Discussion of our observations can also be extended to more fundamental issues, such as the coherence of the superconducting state. The coexistence of a high superconducting transition temperature with such a microscopic inhomogeneity implies that the superconducting coherence length is shorter than the mean free path. Our measured gap correlation length, $\xi \approx 14$ Å, sets the length scale for the superconducting pair size in optimally doped Bi$_2$Sr$_2$CaCu$_2$O$_{8+x}$. By evaluating the BCS expression $\xi_0 = \hbar v_{\mathrm{F}}/\pi\Delta$, taking $\hbar v_{\mathrm{F}} = 1.6$ eVÅ from band dispersion near the nodes [14,16] and the averaged gap at optimal doping as $\Delta = 0.04$ eV, we obtain $\xi_0 \approx 13$ Å, which is in good agreement with the correlation decay length $\xi$ obtained from our experiment. Yet $\xi$ appears to be shorter than the experimental in-plane superconducting coherence length

$\xi_{ab} \approx 22-27\text{Å}$ [17–19]. In contrast to conventional BCS superconductors, it is conceivable that the amplitude and phase coherence in high-$T_c$ superconductors have different length scales, because the ratio $R = 2\Delta/k_B T_c$, is no longer a constant. Recent ARPES measurements [11] suggest that $R \propto 1/x$. Thus we may expect the superconducting phase coherence length to be determined by $\hbar v_F/k_B T_c$, which scales as $1/x$ on the underdoped side. An extension of our correlation length and vortex core-size measurements to underdoped samples with various doping concentrations will perhaps distinguish the two length scales because they may have different doping ($x$) dependences.

The observation of microscopic spatial variations in both the carrier density and the superconducting gap, and the strong correlation between these variations, reveals the local inhomogeneous charge environment in these materials, and its intimate relationship with superconductivity. Further exploration of this frontier may lead to a greater understanding of how high-$T_c$ superconductivity arises from doping a Mott insulator.

We acknowledge P. A. Lee and E. W. Plummer for their comments. We also thank P. W. Anderson, A. Balatsky, D. A. Bonn, A. Castro-Neto, E. Carlson, M. Franz, L. H. Greene, X. Hu, T. Imai, B. Keimer, S. A. Kivelson, K. Kitasawa, R. B. Laughlin, D.-H. Lee, A. H. MacDonald, A. Millis, N. P. Ong, Z.-X. Shen, H.-J. Tao, X.-G. Wen, Z.-Y. Weng, N.-C. Yeh, G.-M. Zhang and Z.-X. Zhong for helpful discussions. This work was supported by the NSF, the DOE, the Sloan Research Fellowship, the Research Corporation, the Miller Institute for Basic Research and a Grant-in-Aid for Scientific Research on Priority Area and a COE Grant from the Ministry of Education, Japan.

[1]  P. W. Anderson, *The Theory of Superconductivity in the High-$T_c$ Cuprates* (Princeton Univ. Press, Princeton, New Jersey, 1997).

[2]  E. W. Hudson, S. H. Pan, A. K. Gupta, K. W. Ng and J. C. Davis, Science **285**, 88 (1999).

[3]  S. H. Pan et al., Nature **403**, 746 (2000).

[4]  Ch. Renner et al., Phys. Rev. Lett. **80**, 149 (1998).

[5]  N. Miyakawa et al., Phys. Rev. Lett. **83**, 1018 (1999).

[6]  H. Hasegawa, H. Ikuta and K. Kitazawa, in *Physical Properties of High Temperature Superconductors* , edited by G. M. Ginsberg (World Scientific, Singapore, 1992).

[7] E. L. Wolf, A. Chang, Z. Y. Rong, Yu. M. Ivanchenko and F. Lu, J. Superconductivity **70**, 355 (1994).

[8] A. Ino *et al.*, Phys. Rev. Lett. **81**, 2124 (1998).

[9] H. F. Fong *et al.*, Nature **398**, 588 (1999).

[10] D. L. Feng *et al.*, Science **289**, 277 (2000).

[11] H. Ding *et al.*, arXiv:cond-mat/0006143.

[12] T. Valla *et al.*, Phys. Rev. Lett. **85**, 828 (2000).

[13] T. Valla *et al.*, Science **285**, 2110 (1999).

[14] A. Kaminski *et al.*, Phys. Rev. Lett. **84**, 1788 (2000).

[15] Y. Zhang *et al.*, Phys. Rev. Lett. **86**, 890 (2001).

[16] P. V. Bogdanov *et al.*, Phys. Rev. Lett. **85**, 2581 (2000).

[17] T. T. M. Palstra *et al.*, Phys. Rev. B **38**, 5102 (1988).

[18] Ch. Renner, B. Revaz, K. Kadowaki, I. Maggio-Aprile and Ø. Fischer, Phys. Rev. Lett. **80**, 3606 (1998).

[19] S. H. Pan *et al.*, Phys. Rev. Lett. **85**, 1536 (2000).

# 9
# Magnetism and Superconductivity in the 122 Family of Iron-Pnictide Superconductors

Y. Gao[1,2], H. X. Huang[2,3], T. Zhou[4], D. G. Zhang[2] and C. S. Ting[2]

[1] *Department of Physics and Institute of Theoretical Physics, Nanjing Normal University, Nanjing, Jiangsu, 210046, China*
[2] *Department of Physics and Texas Center for Superconductivity, University of Houston, Houston, Texas, 77204, USA*
[3] *Department of Physics, Shanghai University, Shanghai, 200444, China*
[4] *College of Science, Nanjing University of Aeronautics and Astronautics, Nanjing 210016, China*

The recently discovered iron-pnictide superconductors exhibit the interplay between magnetism and superconductivity. Understanding how these two phases coexist and compete with each other is of great importance to unraveling the novel properties of these superconductors. Based on a minimum two-orbital tight-binding model which fits the angle resolved photo-emission experiments, we obtain the Fermi surface and phase diagram as a function of the electron doping for the 122 family of the iron-pnictides as observed by experiments. With this background, we review some theoretical aspects, material properties and experimental observations in various doping regimes of these compounds. Hopefully, this study would help us in gaining more insight into the nature of magnetism and superconductivity in the 122 iron-pnictides.

## 9.1   Introduction

The iron-pnictide superconductor LaFeAsO$_{1-x}$F$_x$ with critical temperature 26 K was discovered by Kamihara *et al.* in 2008 [1]. Within two months, materials based on substitution of La with other rare earths had been synthesized, raising the critical temperature of Fe-based materials to 55 K [2]. This rapid sequence of discoveries captured the attention of the high-temperature superconductivity community. In the following three years, the discoveries of several related families of materials were reported. Up to now, the iron-pnictide superconductors can be categorized into three main families, they are the 1111 family ($R$FeAsO, with $R$=La, Ce, Pr, Nd, Sm), 122 family ($A$Fe$_2$As$_2$, with $A$=Ba, Sr, Ca) and 111 family (LiFeAs). Among them, the 122 family, such as the electron-doped Ba(Fe$_{1-x}$Co$_x$)$_2$As$_2$ [3] and the hole-doped Ba$_{1-x}$K$_x$Fe$_2$As$_2$ [4], due to the availability of large homogeneous single crystals, has become one of the most important and most studied materials in iron-pnictide superconductors. The electron-doped 122 system exhibits the coexistence of spin-density-wave (SDW) order [5] and superconductivity, similar to the electron-doped cuprates, providing another promising group of materials for studying the interplay between magnetism and superconductivity. However, unlike the cuprates, the parents compounds are not Mott insulators, but bad metals [5]. Magnetism in these materials is most likely to originate from itinerant electrons and is a result of SDW instability due to Fermi surface (FS) nesting [6]. The phase diagram for these materials [7–11] indicates that the parent compound upon cooling through $T_N \approx 140$ K develops a static collinear SDW order [12]. Increasing the doping of Co or K, the SDW order is suppressed and the superconducting (SC) order emerges as the temperature $T$ falls below $T_c$. The SDW and SC orders coexist in the underdoped samples. By further increasing the Co or K concentration to the optimally doped regime, the SDW order disappears. These experimental results provide compelling evidences for strong competition between the SDW and SC orders.

Another key issue is the SC pairing symmetry. Experimental results about the pairing symmetry in the iron-pnictides remain highly controversial, leaving the perspectives ranging from nodeless to nodal gap structure. Although evidence for a nodal gap has been accumulated in Ba(FeAs$_{1-x}$P$_x$)$_2$ system [13], in the Co- and K-doped 122-family of iron pnictides, the experimental data point to the existence of isotropic gaps, with no node on the FS, especially in the optimally doped samples [14–16]. Theoretically it was suggested that the pairing may be established via inter-pocket scattering of electrons between the hole pockets (around the $\Gamma$ point) and the electron ones (around the $M$ point), leading to the so-called extended $s$-wave or $s_\pm$ pairing symmetry, i.e., $\Delta_k \sim \cos k_x + \cos k_y$ defined in the 2Fe/cell Brillouin zone (BZ) [6], which qualitatively agrees with the experimental observations.

In this review paper, we intend to summarize our group's recent theoretical efforts on iron-pnictide superconductors and show the comparison with experiments. We focused on the magnetism and superconductivity in the 122 family since it has the most available and reliable experimental data. Our studies are based on the Fermi-liquid mean-field (MF) theory, which should be appropriate due to the itinerant nature of the iron-pnictides. The paper is organized as follows: In Sec. 9.2, we introduce our model and explain how it agrees with the angle-resolved photoemission spectroscopy (ARPES) experiments. In Sec. 9.3, we show our results of the phase diagram and FS evolution with doping in $Ba(Fe_{1-x}Co_x)_2As_2$. In Sec. 9.4, we study the asymmetry of the SC coherence peaks in the local density of states (LDOS). In Sec. 9.5, the spin dynamics in $Ba(Fe_{1-x}Co_x)_2As_2$ at different doping levels and temperatures is investigated. In Sec. 9.6, we study the vortex states in $Ba_{1-x}K_xFe_2As_2$. In Sec. 9.7, the domain wall structure is investigated. We point out all of our theoretical studies are consistent with the corresponding experiments. The conclusion is given in Sec. 9.8.

## 9.2   Model and Formalism

Similar to the cuprates, the iron-pnictide superconductors also have layered structures. It has been accepted that superconductivity originates in the Fe-Fe plane. However, in the iron-pnictides, each unit cell contains two Fe ions and two As ions. The four As ions around each Fe ion do not locate in the Fe-Fe plane and have a twofold rotational symmetry and two reflectional symmetries (see Fig. 9.1). Because of different arrays of As ions around Fe ions, the Fe-Fe plane can be divided into two sublattices A and B. The diagonal directions of the Fe-Fe plane have translational symmetry with the period a. In this coordinate system, the momentum is a good quantum number. In the following, based on two Fe ions per unit cell and two degenerate orbitals $d_{xz}$ and $d_{yz}$ per Fe ion, an effective four-band tight-binding model can be constructed which is written as

$$
\begin{aligned}
H_t = -\sum_{\alpha ij\sigma} \{ &\mu(c^{\dagger}_{A\alpha,ij\sigma}c_{A\alpha,ij\sigma} + c^{\dagger}_{B\alpha,ij\sigma}c_{B\alpha,ij\sigma}) \\
&+ [t_1 c^{\dagger}_{A\alpha,ij\sigma}(c_{B\alpha,ij\sigma} + c_{B\alpha,i+1j\sigma} + c_{B\alpha,ij+1\sigma} + c_{B\alpha,i+1j+1\sigma}) \\
&+ t_2(c^{\dagger}_{A\alpha,ij\sigma}c_{A\alpha,i+1j\sigma} + c^{\dagger}_{B\alpha,ij\sigma}c_{B\alpha,ij+1\sigma}) \\
&+ t_3(c^{\dagger}_{A\alpha,ij\sigma}c_{A\alpha,ij+1\sigma} + c^{\dagger}_{B\alpha,ij\sigma}c_{B\alpha,i+1j\sigma}) \\
&+ t_4(c^{\dagger}_{A\alpha,ij\sigma}c_{A\bar\alpha,i+1j\sigma} + c^{\dagger}_{A\alpha,ij\sigma}c_{A\bar\alpha,ij+1\sigma} \\
&+ c^{\dagger}_{B\alpha,ij\sigma}c_{B\bar\alpha,i+1j\sigma} + c^{\dagger}_{B\alpha,ij\sigma}c_{B\bar\alpha,ij+1\sigma}) + h.c.]\},
\end{aligned}
\tag{9.1}
$$

where $c^{\dagger}_{A(B)\alpha,ij\sigma}$ $(c_{A(B)\alpha,ij\sigma})$ creates (destroys) an $\alpha$ electron with spin $\sigma$ in the unit cell $\{i,j\}$ of the sublattice A (B), and $\alpha = 0$ and 1 represent the degenerate orbitals $d_{xz}$ and $d_{yz}$, respectively. Here $\mu$ is the chemical potential, $t_1$ is the hopping between the same orbitals on the nearest-neighbor (NN) Fe sites, $t_2$ and $t_3$ are the next-nearest-neighbor (NNN) hoppings between the same orbitals mediated by As ions B and A, respectively, and $t_4$ is the hopping between the different orbitals on the NNN Fe sites (see Fig. 9.1). It is expected that $t_4$ is small and has the same value along both translationally symmetric directions. In the following, $t_1 = 0.5$, $t_2 = 0.2$, $t_3 = -1$ and $t_4 = 0.02$ (eV) are chosen to fit the FS evolution with doping as observed by ARPES experiments. By diagonalizing Eq. (9.1) in momentum space, four energy bands can be derived and they are shown in Fig. 9.2 along high-symmetry directions in the first BZ. At half filling $\mu = -0.622$ (indicated by the black dashed line in Fig. 9.2), there are two hole pockets ($\alpha$ and $\beta$) around the $\Gamma$ point and two electron ones ($\gamma$ and $\delta$) around the $M$ point, in agreement with ARPES experiments. $\mu < -0.622$ and $\mu > -0.622$ correspond to hole and electron dopings, respectively. The two hole bands $\alpha$ and $\beta$ are not degenerate at the $\Gamma$ point, thus when $\mu > -0.48$, $\alpha$ pocket will disappear first, then $\beta$ pocket will also disappear by further increasing $\mu$ to $-0.32$. The disappearance of the two hole pockets at different doping levels agrees quite well with experimental observations. As can be seen from Fig. 9.3, in electron-doped $BaFe_{2-x}Co_xAs_2$, the two hole pockets around the $\Gamma$ point indeed do not vanish simultaneously with electron doping. When increasing the chemical potential (doping with electrons), $\alpha$ pocket vanishes first, then $\beta$ pocket will vanish by further electron doping. And the inset of Fig. 9.3 clearly shows that $\alpha$ and $\beta$ bands

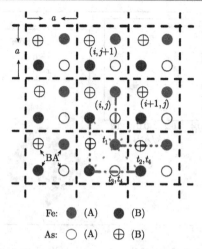

Fe: ● (A)   ● (B)

As: ○ (A)   ⊕ (B)

Fig. 9.1   From [17]. Schematic lattice structure of FeAs layers with each unit cell (denoted by $i$ and $j$) containing two Fe (A and B) and two As (A and B) ions. The As ions A and B are located above and below the center of each Fe square lattice, respectively. Here, $t_1$ is the NN hopping between the same orbitals $d_{xz}$ or $d_{yz}$. $t_2$ and $t_3$ are the NNN hoppings between the same orbitals mediated by the As ions B and A, respectively. $t_4$ is the NNN hopping between the different orbitals.

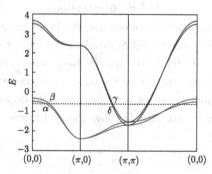

Fig. 9.2   From [17]. The band structure of the four-band tight-binding model plotted along $(0,0) \to (\pi,0) \to (\pi,\pi) \to (0,0)$ in the first BZ. The black dashed line represents $\mu = -0.622$ (half filling).

are not degenerate at the $\Gamma$ point. On the contrary, in the LDA calculated band structure, e.g., see Fig. 1(b) in [19], close to the Fermi energy, the two hole bands are degenerate at the $\Gamma$ point. Thus when increasing the chemical

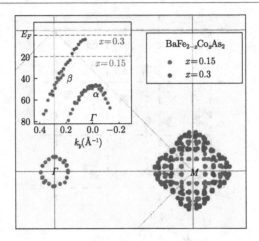

Fig. 9.3   From [18]. Comparison of the experimentally determined $k_F$ points between BaFe$_{1.7}$Co$_{0.3}$As$_2$ and BaFe$_{1.85}$Co$_{0.15}$As$_2$. The inset shows the experimental band dispersion in the vicinity of $E_F$ around the $\Gamma$ point. The chemical potential of the Co$_{0.3}$ sample is shifted upward by 20 meV with respect to that of the Co$_{0.15}$ sample.

potential, the two hole pockets around the $\Gamma$ point will vanish simultaneously, in apparent contradiction to ARPES observations.

From the above comparison we can see, the band structures calculated by LDA and those tight-binding models fitted to the LDA results are all inconsistent with experiments while the energy band structure described by our tight-binding model agrees qualitatively with the observations of ARPES experiments in the whole range of electron and hole dopings, indicating the validity of our model in studying the low-energy physics of iron-pnictide superconductors.

## 9.3   Phase Diagram and Fermi Surface Evolution

Based on the above tight-binding model, we then take the SDW and SC orders into account. The effective model includes the on-site interactions which are solely responsible for the SDW and the NNN intraorbital attraction which causes the $s_\pm$ paring. It can be written as

$$H = H_0 + H_\Delta + H_{int}. \tag{9.2}$$

The first term is the hopping term suggested by Eq. (9.1) and can be expressed as

$$H_0 = - \sum_{i\mu j\nu\sigma} (t_{i\mu j\nu} c_{i\mu\sigma}^\dagger c_{j\nu\sigma} + h.c.) - t_0 \sum_{i\mu\sigma} c_{i\mu\sigma}^\dagger c_{i\mu\sigma}, \tag{9.3}$$

where $i, j$ are the site indices and $\mu, \nu = 1, 2$ are the orbital indices. $t_0$ is the chemical potential.

$H_\Delta$ is the pairing term,

$$H_\Delta = \sum_{i\mu j\nu\sigma} (\Delta_{i\mu j\nu} c_{i\mu\sigma}^\dagger c_{j\nu\bar{\sigma}}^\dagger + h.c.). \tag{9.4}$$

$H_{int}$ is the on-site interaction term. Following [20], we here include the intra- and inter-orbital Coulombic interactions $U$ and $U'$, as well as the Hund's coupling $J_H$. At the MF level, the interaction Hamiltonian can be written as [21]

$$H_{int} = U \sum_{i\mu\sigma\neq\bar{\sigma}} \langle n_{i\mu\bar{\sigma}} \rangle n_{i\mu\sigma} + U' \sum_{i,\mu\neq\nu,\sigma\neq\bar{\sigma}} \langle n_{i\mu\bar{\sigma}} \rangle n_{i\nu\sigma}$$
$$+ (U' - J_H) \sum_{i,\mu\neq\nu,\sigma} \langle n_{i\mu\sigma} \rangle n_{i\nu\sigma}, \tag{9.5}$$

where $n_{i\mu\sigma}$ is the density operator at the site $i$ and orbital $\mu$, with spin $\sigma$. $U'$ is taken to be $U - 2J_H$ [20].

Then the Hamiltonian can be diagonalized by solving the BdG equations self-consistently,

$$\sum_j \sum_\nu \begin{pmatrix} H_{i\mu j\nu\sigma} & \Delta_{i\mu j\nu} \\ \Delta_{i\mu j\nu}^* & -H_{i\mu j\nu\bar{\sigma}}^* \end{pmatrix} \begin{pmatrix} u_{j\nu\sigma}^n \\ v_{j\nu\bar{\sigma}}^n \end{pmatrix} = E_n \begin{pmatrix} u_{i\mu\sigma}^n \\ v_{i\mu\bar{\sigma}}^n \end{pmatrix}, \tag{9.6}$$

where $H_{i\mu j\nu\sigma}$ is expressed by

$$H_{i\mu j\nu\sigma} = -t_{i\mu j\nu} + [U\langle n_{i\mu\bar{\sigma}} \rangle + (U - 2J_H)\langle n_{i\bar{\mu}\bar{\sigma}} \rangle$$
$$+ (U - 3J_H)\langle n_{i\bar{\mu}\sigma} \rangle - t_0]\delta_{ij}\delta_{\mu\nu}. \tag{9.7}$$

The SC order parameter and the local electron density $\langle n_{i\mu} \rangle$ satisfy the following self-consistent conditions,

$$\Delta_{i\mu j\nu} = \frac{V_{i\mu j\nu}}{4} \sum_n (u_{i\mu\uparrow}^n v_{j\nu\downarrow}^{n*} + u_{j\nu\uparrow}^n v_{i\mu\downarrow}^{n*}) \tanh\left(\frac{E_n}{2k_B T}\right), \tag{9.8}$$

$$\langle n_{i\mu} \rangle = \sum_n |u_{i\mu\uparrow}^n|^2 f(E_n) + \sum_n |v_{i\mu\downarrow}^n|^2 [1 - f(E_n)]. \tag{9.9}$$

Here $V_{i\mu j\nu}$ is the pairing strength and $f(x)$ is the Fermi distribution function. The LDOS is expressed by

$$\rho_i(\omega) = \sum_{n\mu} [|u_{i\mu\sigma}^n|^2 \delta(E_n - \omega) + |v_{i\mu\bar{\sigma}}^n|^2 \delta(E_n + \omega)]. \tag{9.10}$$

The supercell technique [22] is used to calculate the LDOS.

The hopping constants are written as

$$t_{i\mu,i\pm\hat{\alpha}\mu} = t_1 \qquad (\alpha = \hat{x}, \hat{y}), \tag{9.11}$$

$$t_{i\mu,i\pm(\hat{x}+\hat{y})\mu} = \frac{1 + (-1)^i}{2} t_2 + \frac{1 - (-1)^i}{2} t_3, \tag{9.12}$$

$$t_{i\mu,i\pm(\hat{x}-\hat{y})\mu} = \frac{1 + (-1)^i}{2} t_3 + \frac{1 - (-1)^i}{2} t_2, \tag{9.13}$$

$$t_{i\mu,i\pm\hat{x}\pm\hat{y}\nu} = t_4 \qquad (\mu \neq \nu), \tag{9.14}$$

here, $\hat{x}$ and $\hat{y}$ are the coordinates connecting the NN Fe ions.

The pairing symmetry is determined by the pairing potential $V_{i\mu j\nu}$. We have carried out extensive calculations to search for favorable pairing symmetries based on the present band structure. Especially, the NNN intra-orbital potential will produce the $s_\pm$ pairing symmetry. Here the pairing symmetry we obtained is independent of the initial input values and the result is consistent with previous calculation based on a different two-orbital model [21]. Also, it agrees with the theoretically proposed pairing symmetry based on spin fluctuations [6] and is qualitatively consistent with ARPES measurement [15]. In the following calculation we focus on this pairing symmetry and the energy is measured in units of $t_1$.

In Fig. 9.4(a), we plot the spatial distribution of the magnetic order [$m_i = \frac{1}{4}\sum_\mu(n_{i\mu\uparrow} - n_{i\mu\downarrow})$] at zero temperature and zero doping. As seen, the magnetic order is antiferromagnetic along the $x$ direction and ferromagnetic along the $y$ direction, corresponding to the $(\pi, 0)$ [$(\pi, \pi)$] SDW in the 1Fe/cell [2Fe/cell] BZ. This result is consistent with the neutron scattering (NS) experiment [12] and previous theoretical calculation based on different band structure [21]. There exists another degenerate SDW state with the spin order being antiferromagnetic along the $y$ direction and ferromagnetic along the $x$ direction.

The fourfold symmetry breaking and the presence of the SDW order are due to the FS nesting at low doping which will be discussed below. The magnetic order decreases as the temperature or doping increases. The magnitudes of the SC and magnetic orders as functions of the temperature and doping are shown in Figs. 9.4(b) and 9.4(c), respectively. At a fixed doping, both the magnetic and SC orders decrease as the temperature increases and two transition temperatures are revealed, as seen in Fig. 9.4(b). The superconductivity occurs at the doping 0.01 and vanishes at the doping 0.26, as seen in Fig. 9.4(c). The magnetic order is maximum at zero doping and decreases monotonically as the doping increases. The calculated phase diagram is plotted in Fig. 9.4(d). As seen, the magnetic and SC orders coexist in the underdoped region. The magnetic order decreases abruptly and a quantum critical point $\delta = 0.12$ is revealed. The superconductivity appears as the magnetic

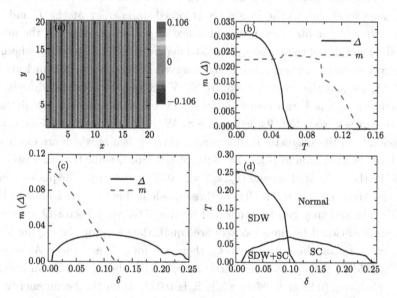

Fig. 9.4   From [23]. (a) The intensity plot of the magnetic order at zero doping and zero temperature. (b) The magnitude of the SC order parameter $\Delta$ and magnetic order $m$ as a function of the temperature with the doping density $\delta = 0.09$. (c) The magnitude of the SC order parameter $\Delta$ and magnetic order $m$ as a function of the doping at zero temperature. (d) The calculated phase diagram.

order is suppressed and the SC transition temperature $T_c$ reaches the maximum as the magnetic order disappears. Our results are reasonably consistent with the experiments in $Ba(Fe_{1-x}Co_x)_2As_2$ [7–9].

After presenting the phase diagram, we then want to investigate the low-energy properties of the iron-pnictide superconductors. Since the low-energy physics is expected to have an intimate relation with the FS topology, we need to study the FS evolution with doping.

At zero doping, below the SDW transition temperature $T_N$, it was proposed experimentally that there exist small FSs along the $\Gamma$-$M$ line of the BZ and Dirac cones in the electronic structure form inside these FSs, with their apices being located close to the Fermi energy. However, whether these FSs and Dirac cones are electron- or hole-like is within uncertainties of the experiment [24]. On the other hand, theoretically it was proposed that in both a two-band model and a five-band model, nodes in the SDW gap function must exist due to the symmetry-enforced degeneracy at the $\Gamma$ and $M$ high-symmetry points, even in the presence of perfect nesting, but the number and locations of these nodes are model dependant [25]. Therefore, whether they correspond to the experimentally observed Dirac cones is still unclear. In Fig. 9.5, we plot the zero temperature SDW FS and the corresponding band structure near the Fermi energy obtained by our self-consistent calculation. As we can see from Fig. 9.5(a), in the SDW state, there remain four small FS pockets in the magnetic Brillouin zone (MBZ), two of which are electron-like (red) located around $(k_x, k_y) = \pm(0.286\pi, 0.286\pi)$, while the other two are hole-like (blue) located around $(k_x, k_y) = \pm(0.308\pi, 0.308\pi)$. The pockets outside the MBZ are just the replica of those inside it due to band-folding in the SDW state and they can be connected by the SDW wave vector $Q = (\pi, \pi)$. The areas enclosed by these pockets are equal, thus keeping the doping level at $x = 0$. Inside these four pockets, there are four Dirac cones. As shown in Fig. 9.5(b), the apex of the Dirac cone is 0.026 below the Fermi energy at $(k_x, k_y) = \pm(0.286\pi, 0.286\pi)$ while it is 0.046 above the Fermi energy at $(k_x, k_y) = \pm(0.308\pi, 0.308\pi)$, suggesting that they are electron- and hole-like Dirac cones, respectively. The spectral function $A(k, \omega)$, which is proportional to the photoemission intensity measured in ARPES experiments, is integrated from $\omega = -0.1$ to $\omega = 0.1$ and shown in Fig. 9.5(c). As we can see, the locations of the bright spots are around $(k_x, k_y) = \pm(0.3\pi, 0.3\pi)$ and the equivalent symmetry points outside the MBZ, on the $\Gamma$-$M$ line, in qualitative agreement with experiment [24]. In addition, although most parts of the original FSs around $\Gamma$ are gapped by the SDW order, the gap value is extremely small on these FSs. Thus, around $\Gamma$, the low-energy spectral function has moderate intensity and this can be seen from the ring structure around $\Gamma$ with lower intensity, as compared to those bright spots. We also notice that the system has only twofold symmetry when entering the SDW state while the experimentally

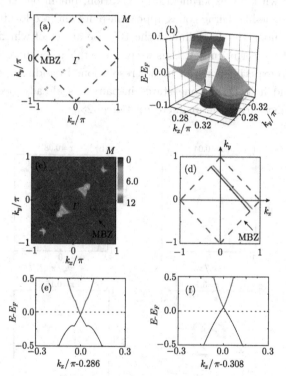

Fig. 9.5   From [26]. (a) The zero temperature SDW FS. (b) Two Dirac cones at $(k_x, k_y) = (0.286\pi, 0.286\pi)$ and $(k_x, k_y) = (0.308\pi, 0.308\pi)$, respectively. (c) The spectral function $A(\mathbf{k}, \omega)$ integrated from $\omega = -0.1$ to $\omega = 0.1$. (e) and (f) are the band structures near the Fermi energy along the blue [goes through $(k_x, k_y) = (0.286\pi, 0.286\pi)$] and orange [goes through $(k_x, k_y) = (0.308\pi, 0.308\pi)$] lines in (d), respectively. The BZ is defined in the 2Fe/cell representation and the green dashed line in (a), (c) and (d) represents the MBZ.

observed fourfold symmetry is due to the superposition of twin domains or domain averaging, as suggested in [24] and [27] respectively. The band structures
near the Fermi energy scanned along the blue and orange cuts in Fig. 9.5(d)
are plotted in Figs. 9.5(e) and 9.5(f), respectively. It clearly shows the X-like
structure of Dirac cones and again suggests that the Dirac cone is electron-
like at $(k_x, k_y) = (0.286\pi, 0.286\pi)$ and hole-like at $(k_x, k_y) = (0.308\pi, 0.308\pi)$.
The locations of the FS pockets and the bright spots in the spectral function
are consistent with the experimental observation, but in our calculation, the
electron- and hole-like Dirac cones appear in-pairs and are located very close
to each other along the $\Gamma$-$M$ line of the BZ, the apices of which are both in
the vicinity of the Fermi energy, thus we propose this to be directly verified by
future ARPES experiments with higher resolution. In addition, the existence
of electron and hole Dirac cone states in-pairs has already been confirmed
indirectly by measuring the magnetoresistance [28].

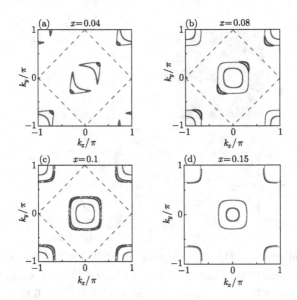

Fig. 9.6  From [26]. The zero temperature SDW FS at various doping levels. The
SC order $\Delta$ is artificially set to zero in order to illustrate the effect of SDW on the
evolution of the FS. The blue and red pockets in the $x = 0.04$, 0.08 and 0.1 cases are
both electron pockets. The green dashed line is the same as that in Fig. 9.5.

Figure 9.6 shows the evolution of the FS with doping. Here, we set the SC order $\Delta$ to zero to illustrate the effect of SDW on the evolution of the FS. In the MBZ, as doping increases, the size of the electron pockets [the red pockets shown in Fig. 9.5(a)] is enlarged while that of the hole pockets [the blue pockets shown in Fig. 9.5(a)] is reduced. When doping increases to about $x = 0.02$, the hole pockets vanish completely. By further increasing doping, another two electron pockets appear in the MBZ [the blue pockets in the $x = 0.04$, 0.08 and 0.1 cases shown in Fig. 9.6], exactly at the same locations where the hole pockets vanish and overlap with the original electron pockets. The size of all these electron pockets is enlarged with doping. If we define the areas enclosed by the inner and outer red lines to be $S_1$ and those enclosed between the inner and outer blue lines to be $S_2$, then we have $x = 2N_x N_y (S_1 + S_2)$, with $N_x$, $N_y$ being the linear dimensions of the square lattice. Finally, when $x = 0.15$, the SDW order disappears and there is no more band-folding due to it. In this case, there are two electron pockets and two hole pockets around the $M$ and $\Gamma$ points of the BZ, respectively.

## 9.4    Asymmetry of the Superconducting Coherence Peaks in the Local Density of States

We also calculate the LDOS spectra according to Eq. (9.10). The spectra at different doping densities are shown in Fig. 9.7. At zero doping [Fig. 9.7(a)], the density of states shows four well-defined coherence peaks due to the SDW order. The maximum dip of the spectrum is at the chemical potential or zero energy. Note that there are two coherence peaks at negative energies and two at positive energies. The splitting of the coherence peaks is caused by the inter-orbital coupling $t_4$. As the doping increases [Figs. 9.7(b) and 9.7(c)], the maximum dip of the LDOS due to the $(\pi, \pi)$ SDW order is nearly independent of doping and always pinned at the chemical potential of zero doping. This is similar to the case of $d$-density wave in cuprate superconductors [29]. As we define the zero energy at the chemical potential, the maximum dip of the SDW spectrum will shift to the left or the negative energy as doping increases. While in the SC state, two SC coherence peaks show up, and the mid-gap point is always located at the zero energy or the chemical potential of the electron-doped system.

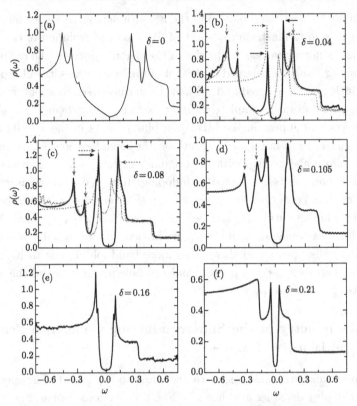

Fig. 9.7   From [23]. The LDOS spectra with different doping densities at the zero temperature. The (red) dash-dotted and (blue) dotted lines in panels (b) and (c) are nonself-consistent results with $\Delta \equiv 0$, and $m \equiv 0$, respectively.

We now discuss the LDOS spectra in the coexisting region. In this region, the pure SC state and the pure SDW state have higher free energy. These two pure states can be obtained by setting the magnetic order $m \equiv 0$ while the SC order is calculated self-consistently and setting the SC order $\Delta \equiv 0$ while the magnetic order is obtained self-consistently, respectively. The results of the pure SC and SDW states are represented by (blue) dotted and (red) dash-dotted lines, respectively in Figs. 9.7(b) and 9.7(c). In the pure SDW state, four SDW peaks are obtained with the maximum dip at the negative energy. In the pure SC state, two SC coherence peaks are seen with the mid point of the gap at the zero energy. In the coexisting state, the structure of the spectra inside the SC gap is practically the same to the spectra of the pure SC state. As shown in Figs. 9.7(b) and 9.7(c), we can see clearly the previous SDW peaks outside the SC gap. But the SDW peaks inside the SC gap in the pure magnetic state cannot be seen in the coexisting state. The additional peaks outside the SC peaks, positioned by the (red) vertical arrows can be regarded as one signature of the coexistence of the magnetic and SC orders in electron-doped materials. As doping increases, only negative SDW peaks would appear, as seen in Figs. 9.7(c) and 9.7(d). It is needed to point out that such structures outside the SC gap due to the SDW order as shown in Figs. 9.7(a–d) so far have not been clearly identified by the experiments.

Another significant feature caused by the magnetic order can be seen from the intensities of the SC coherence peaks [see Figs. 9.7(b) and 9.7(c)]. In pure SC state, the intensity of the coherent SC peak at the negative energy is higher than that at the positive energy for all the doping densities we considered. These SC peaks are positioned by the parallel (blue) dotted arrows. In the coexisting region, as we mentioned above, the SDW spectra shift to the negative energy so that the SC peak at the negative energy is within the SDW gap and that at the positive energy is outside the SDW gap. Thus the intensity of the SC peak at the positive energy is enhanced and the one at the negative energy gets suppressed by the SDW order. As a result, in the coexisting state, the intensity of the SC coherence peak [denoted by the (black) solid arrows] at the negative energy is lower than that at the positive energy. The asymmetry disappears near the optimal doping ($\approx 0.105$), as seen in Fig. 9.7(d). In the overdoped region [see Figs. 9.7(e) and 9.7(f)], the asymmetry occurs again but with the intensity of the SC peak at negative energy becoming stronger. This feature has recently been confirmed by the scanning tunneling microscopy (STM) experiments on $BaFe_{2-x}Co_xAs_2$ [30].

## 9.5  Spin Dynamics

Recently, several NS experiments have been carried out to probe the spin dynamics in the iron-pnictides [7,9,31–38], and the spin excitation spectrum was fitted by using a $J_1, J_2$ Heisenberg model based on localized spins [31,32,38,39]. However, while the parent compounds of the cuprates are Mott insulators with

large in-plane exchange interactions [40], the parent iron-pnictides are bad
metals and remain itinerant at all doping levels. Magnetism in these materials
are most likely to originate from itinerant electrons and is a result of SDW
instability due to FS nesting [6]. Theoretically, at present, the variation of the
spin susceptibility with doping remains less explored. The spin susceptibilities
were mostly studied in the optimally doped compounds without SDW [19,41]
or in the parent compound without superconductivity [42], as well as in the
normal state with neither SDW nor superconductivity [43]. Thus, we present a
comprehensive study of the spin dynamics with doping in $Ba(Fe_{1-x}Co_x)_2As_2$,
by taking the interplay between SDW and SC orders into full account. We
adopt Fermi-liquid MF theory to study the static SDW and superconductivity,
and employ the random-phase approximation (RPA) to investigate the spin
dynamics from the imaginary part of the dynamic spin susceptibility.

The Hamiltonian of our system can be expressed as

$$H = H_0 + H_\Delta + H_{int}. \tag{9.15}$$

$H_0$, $H_\Delta$ and $H_{int}$ are the momentum-space transformations of Eqs. (9.3), (9.4)
and (9.5), respectively. They can be written as

$$H_0 = \sum_{k\sigma} \psi_{k\sigma}^\dagger M_k \psi_{k\sigma}, \tag{9.16}$$

where $\psi_{k\sigma}^\dagger = (c_{A0,k\sigma}^\dagger, c_{A1,k\sigma}^\dagger, c_{B0,k\sigma}^\dagger, c_{B1,k\sigma}^\dagger)$, and

$$M_k = \begin{pmatrix} \varepsilon_{A,k} - \mu & \varepsilon_{xy,k} & \varepsilon_{T,k} & 0 \\ \varepsilon_{xy,k} & \varepsilon_{A,k} - \mu & 0 & \varepsilon_{T,k} \\ \varepsilon_{T,k} & 0 & \varepsilon_{B,k} - \mu & \varepsilon_{xy,k} \\ 0 & \varepsilon_{T,k} & \varepsilon_{xy,k} & \varepsilon_{B,k} - \mu \end{pmatrix}, \tag{9.17}$$

with

$$\varepsilon_{A,k} = -2(t_2 \cos k_x + t_3 \cos k_y),$$

$$\varepsilon_{B,k} = -2(t_2 \cos k_y + t_3 \cos k_x),$$

$$\varepsilon_{xy,k} = -2t_4(\cos k_x + \cos k_x),$$

$$\varepsilon_{T,k} = -4t_1 \cos \frac{k_x}{2} \cos \frac{k_y}{2}. \tag{9.18}$$

Here the momentum $k$ is defined in the 2Fe/cell BZ.

The pairing term is

$$H_\Delta = \sum_{ks\alpha}(\Delta_k c^\dagger_{s\alpha,k\uparrow} c^\dagger_{s\alpha,-k\downarrow} + h.c.). \tag{9.19}$$

Here, $s = 0$ $(A)$ or $1$ $(B)$ is the sublattice index, and the SC order parameter is $\Delta_k = \dfrac{\Delta_0}{2}(\cos k_x + \cos k_y)$, where

$$\Delta_0 = \frac{2V_{nnn}}{N}\sum_k (\cos k_x + \cos k_y)\langle c_{s\alpha,-k\downarrow}c_{s\alpha,k\uparrow}\rangle, \tag{9.20}$$

with $V_{nnn}$ being the NNN attractive pairing interaction and $N$ being the number of unit cells.

Assuming the $(\pi, \pi)$ SDW order in the 2Fe/cell BZ, in momentum space, $H_{int}$ can be expressed as

$$H_{int} = \frac{n}{4}(3U - 5J_H)\sum_{ks\alpha\sigma} n_{s\alpha,k\sigma} - \frac{m}{2}(U + J_H)\sum_{ks\alpha\sigma} \sigma c^\dagger_{s\alpha,k+Q\sigma}c_{s\alpha,k\sigma}, \tag{9.21}$$

where $n = 2 + x$ is the number of electrons per lattice site and $\boldsymbol{Q} = (\pi, \pi)$. The SDW order parameter is

$$m = \frac{1}{N}\sum_{k\sigma} \sigma\langle c^\dagger_{s\alpha,k+Q\sigma}c_{s\alpha,k\sigma}\rangle. \tag{9.22}$$

The effective momentum-space Hamiltonian is then given by

$$H = {\sum_k}' \varphi^\dagger_k W_k \varphi_k,$$

$$W_k = \begin{pmatrix} M'_k & R & \Delta_k I & 0 \\ R & M'_{k+Q} & 0 & -\Delta_k I \\ \Delta_k I & 0 & -M'_k & R \\ 0 & -\Delta_k I & R & -M'_{k+Q} \end{pmatrix}, \tag{9.23}$$

where

$$\varphi^\dagger_k = (\psi^\dagger_{k\uparrow}, \psi^\dagger_{k+Q\uparrow}, \psi_{-k\downarrow}, \psi_{-(k+Q)\downarrow}),$$

$$M'_k = M_k + \frac{n}{4}(3U - 5J_H)I, \tag{9.24}$$

and

$$R = \begin{cases} -\dfrac{m}{2}(U + J_H)J, & \text{when } k \text{ is located in the first or third quadrant,} \\ -\dfrac{m}{2}(U + J_H)I, & \text{when } k \text{ is located in the second or fourth quadrant,} \end{cases}$$

$$Q = \begin{cases} (\pi, \pi), & \text{when } k \text{ is located in the first or third quadrant,} \\ (-\pi, \pi), & \text{when } k \text{ is located in the second or fourth quadrant.} \end{cases} \tag{9.25}$$

$I$ is a $4 \times 4$ unit matrix, $J = diag\{-1, -1, 1, 1\}$ and $\sum\limits_{k}'$ means the summation extends over the MBZ: $-\pi < k_x \pm k_y \leqslant \pi$. The MF Green's function matrix can be written as

$$g(\boldsymbol{k}, \tau) = -\langle T_\tau \varphi_{\boldsymbol{k}}(\tau) \varphi_{\boldsymbol{k}}^\dagger(0) \rangle, \tag{9.26}$$

and

$$g(\boldsymbol{k}, ip_n) = A_{\boldsymbol{k}} W_{\boldsymbol{k}}' A_{\boldsymbol{k}}^\dagger, \tag{9.27}$$

where $W_{\boldsymbol{k}ij}' = \delta_{ij}(ip_n - \lambda_{\boldsymbol{k}i})^{-1}$ and $A_{\boldsymbol{k}}$ is a unitary matrix that satisfies $(A_{\boldsymbol{k}}^\dagger W_{\boldsymbol{k}} A_{\boldsymbol{k}})_{ij} = \delta_{ij}\lambda_{\boldsymbol{k}i}$.

The MF spin susceptibility is

$$\chi_{t\gamma,u\delta}^{r\alpha,s\beta(0)}(\boldsymbol{q}, \boldsymbol{q}', i\omega_n) = \frac{\delta_{\boldsymbol{q}',\boldsymbol{q}}}{2} \sum_{(i,j)=(p,o)}^{(o+8,p+8)} [P_{im,nj}(q) + P_{i+4m,nj+4}(q)], \tag{9.28}$$

here, $r, s, t, u$ label the sublattice indices, $\alpha, \beta, \gamma, \delta$ represent the orbitals, with

$$m = 2r + \alpha + 1,$$
$$n = 2s + \beta + 1,$$
$$o = 2t + \gamma + 1,$$
$$p = 2u + \delta + 1, \tag{9.29}$$

and

$$P_{im,nj}(q) = -\frac{1}{\beta N} \sum_{k} g_{im}(k) g_{nj}(k + q). \tag{9.30}$$

Here we use $k = (\boldsymbol{k}, ip_n)$ and $q = (\boldsymbol{q}, i\omega_n)$.

We then use RPA to take into account the residual fluctuation of $H_{int}$ beyond MF. The RPA spin susceptibility is determined by the matrix equation

$$\chi^{RPA}(q) = \sum_{rt\alpha\gamma} \{\chi^0(q)[I - \Gamma\chi^0(q)]^{-1}\}_{t\gamma,t\gamma}^{r\alpha,r\alpha}, \tag{9.31}$$

where $I$ is a $16 \times 16$ unit matrix and the nonzero elements of the interaction vertex are: for $\alpha = \beta = \gamma = \delta$, $\Gamma_{r\gamma,r\delta}^{r\alpha,r\beta} = 2U$; for $\alpha = \beta \neq \gamma = \delta$ or $\alpha = \gamma \neq \beta = \delta$, $\Gamma_{r\gamma,r\delta}^{r\alpha,r\beta} = 2J_H$.

First, we solve the MF equations self-consistently to obtain $m$, $\Delta_0$ and $\mu$ at different doping levels $x$ and temperatures $T$. The calculated phase diagram as shown in the inset of Fig. 9.8(a) reproduces the result based on Bogoliubov-de Gennes (BdG) equations [see Fig. 9.4(d)] and is also qualitatively consistent with the experiments on Ba(Fe$_{1-x}$Co$_x$)$_2$As$_2$ [7–9]. Here the SDW and SC orders are competing with each other. If there is no SDW, the SC order would show up even in the parent compound. The presence of the SC order also suppresses SDW. For example, in the underdoped ($x = 0.06$) compound with $T_N \approx 0.16$ and $T_c \approx 0.05$, the calculated magnitude of $m^2$ (proportional to the magnetic Bragg peak intensity) at $T = 0$ is reduced by $\sim 4\%$ relative to that of the maximum intensity at $T_c$ [see Fig. 9.8(a)], and this result is consistent with the neutron diffraction experiments [7,31].

Then, we investigate the spin dynamics in Ba(Fe$_{1-x}$Co$_x$)$_2$As$_2$ for $x = 0$, 0.06, 0.1, and 0.2, corresponding to the undoped, underdoped, optimally doped and overdoped compounds, respectively.

In the parent ($x=0$) compound, the RPA spin susceptibility [see Fig. 9.8(b)] in the paramagnetic state at $T = 0.27$ ($T_N \approx 0.25$) shows a linear energy dependence for $\omega < 0.1$, suggesting gapless excitations [33,37]. On the other hand, in the SDW state at $T = 0$, the spin excitation intensity is close to zero below $\omega \approx 0.13$, similar to a spin gap [32,33]. However, the gap is not sharp since a sharp gap would produce a stepwise increase in intensity at the gap energy which is unlike the more gradual increase seen here. Figures 9.8(c) and 9.8(d) show Im$\chi^{RPA}(\boldsymbol{q}, \omega)$ as a function of energy transfer $\omega$ and momentum $\boldsymbol{q}$ along $(q_x/\pi, q_y/\pi) = (h, h)$ direction at $T = 0$. As we can see, there is almost no detectable intensity below $\omega \approx 0.13$, again illustrating the opening of the spin gap. The excitations are peaked at $\boldsymbol{Q} = (\pi, \pi)$ and $\omega \approx 0.93$, at higher energies, the response is seen to split and broaden due to the dispersion of the spin waves. By tracking the peak positions in Fig. 9.8(c), the spin-wave dispersion relation can be fitted as $\omega_q = \sqrt{\Delta^2 + v^2 q^2}$ [33,38,45], where $\Delta \approx 0.93$ is an energy gap, $v \approx 9.02$ is the spin-wave velocity, and $q$ is the reduced wave vector away from $(1, 1)$ along the $(h, h)$ direction. Here we need to make it clear that $\omega \approx 0.13$ is a spin gap, since below it, there is neither spin-flip

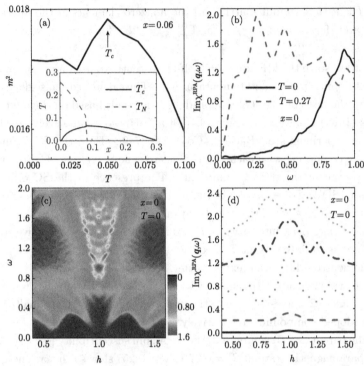

Fig. 9.8   From [44]. (a) $m^2$ as a function of $T$ close to $T_c$ at $x = 0.06$. Inset shows the calculated phase diagram. (b) $\mathrm{Im}\chi^{RPA}(\boldsymbol{q}, \omega)$ at $\boldsymbol{q} = (\pi, \pi)$ as a function of $\omega$ by changing $i\omega_n$ to $\omega + i\eta$, at $x = 0$ and different temperatures $T$. (c) $\mathrm{Im}\chi^{RPA}(\boldsymbol{q}, \omega)$ as a function of energy transfer $\omega$ and momentum $\boldsymbol{q}$, at $x = 0$ and $T = 0$. The momentum is scanned along $(q_x/\pi, q_y/\pi) = (h, h)$. (d) Constant-energy scans along the $(q_x/\pi, q_y/\pi) = (h, h)$ direction at $x = 0$ and $T = 0$. Successive cuts are displaced vertically for clarity. The energy transfer is $\omega = 0.09$ (black solid), 0.4 (red dash), 0.93 (green dot), 1.2 (blue dash dot), and 2 (cyan short dash). The damping rate $\eta = 0.04$.

particle-hole excitation nor collective spin wave excitation. From $\omega \approx 0.13$ to $\omega \approx 0.93$, there is spin-flip particle-hole excitation, but no spin wave excitation. Only above $\omega \approx 0.93$, there is collective spin wave excitation and thus $\Delta \approx 0.93$ is an energy gap in the spin wave spectrum. Experimentally observed gap (7.7 meV or 9.8 meV, see the inset of Fig. 3 in [32] and Fig. 3(a) in [33]) refers to the region where there is neither spin-flip particle-hole excitation nor collective spin wave excitation, corresponding to the $\omega \approx 0.13$ case in our paper. The origin of the spin gap can be understood in terms of the Fermi surface at $x = 0$ as shown in Fig. 9.5(a). In the paramagnetic state, large parts of the two hole pockets around $\Gamma = (0,0)$ and two electron pockets around $M = (\pi, \pi)$ are nested by momentum $(\pi, \pi)$, thus giving rise to the gapless excitations at $T = 0.27$. But at $T = 0$, the SDW order will gap most parts of the original Fermi surface, leaving only tiny ungapped Fermi surface pockets connected by $(\pi, \pi)$ along the $\Gamma - M$ line, so in this case, for small energies, the imaginary part of the MF spin susceptibility is close to zero, while its real part does not fulfill the resonance condition, leading to a spin gap opening in the RPA spin susceptibility.

The RPA spin susceptibility [Fig. 9.9(a)] at $x = 0.06$ suggests that the excitations above $T_c$ are gapless, although the intensity is very small at low energy. This may be the reason why above $T_c$, Ref. [7] claims the excitations are gapless while Ref. [31] concludes they are gapped. Below $T_c$, the intensities below $\omega \approx 0.06$ and above $\omega \approx 0.36$ are suppressed and the weight is transferred to form a resonance at $\omega_{res} \approx 0.14$. Since $m(T = 0.09) \approx 0.129$ and $m(T = 0) \approx 0.131$, our results seem to agree with Ref. [7], which claims the resonance is produced by suppressing low energy spectral weight, rather than Ref. [31], where the spectral weight is considered to be transferred from the ordered magnetic moments. In addition, Fig. 9.9(b) shows that below $\omega_{res}$, commensurate spin excitation prevails, in agreement with experimental observation [31], and it becomes incommensurate when the energy is above $\omega_{res}$, notably between $\omega \approx 0.15$ and $\omega \approx 0.18$, which we predict to be measurable by NS experiment. The spin excitations can extend beyond $\omega = 1$, with smeared out and broadened features for $\omega \gtrsim 0.2$.

At $x = 0.1$, the SDW order is completely suppressed, and SC emerges for $T < T_c \approx 0.06$. The excitation spectrum [Fig. 9.9(c)] shows that in the superconducting state at $T = 0$, a gap below $\omega \approx 0.08$ develops and there is a resonance above the gap energy peaking at $\omega_{res} \approx 0.24$, qualitatively agrees with the NS experiments on the optimally doped $Ba(Fe_{1-x}Co_x)_2As_2$ [34–37]. Furthermore, Fig. 9.9(d) shows that the spin excitation is incommensurate at low energy $(0.18 \lesssim \omega \lesssim \omega_{res})$ which still need to be verified by experiments, then it switches to a commensurate behavior between $\omega \approx \omega_{res}$ and $\omega \approx 0.3$, and becomes broad at higher energy, consistent with Refs. [35,36,38]. In the

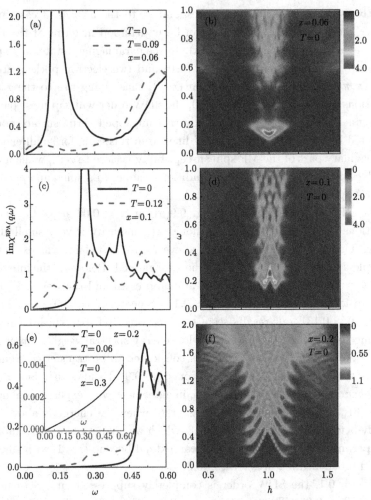

Fig. 9.9   From [44]. (a) $\mathrm{Im}\chi^{RPA}(\boldsymbol{q},\omega)$ at $\boldsymbol{q}=(\pi,\pi)$ as a function of $\omega$, at $x=0.06$ and different temperatures $T$. (b) $\mathrm{Im}\chi^{RPA}(\boldsymbol{q},\omega)$ as a function of energy transfer $\omega$ and momentum $\boldsymbol{q}$, at $x=0.06$ and $T=0$. The momentum is scanned along $(q_x/\pi, q_y/\pi)=(h,h)$. (c) and (d) [(e) and (f)] are similar to (a) and (b), respectively, but at $x=0.1$ ($x=0.2$). Inset in (e) shows the $x=0.3$ case.

normal state at $T=0.12$, the spectrum is replaced by broad gapless excitations with a linear energy dependence for $\omega < 0.1$ [36,37]. We notice a marked similarity between the spin excitations in the normal state of the optimally doped compound and those in the paramagnetic state of the parent compound as observed in [37], suggesting a common origin of spin fluctuations in both of them.

In contrast, the spin excitations in the overdoped ($x = 0.2$) compound [Fig. 9.9(e)] show gap-like behavior in both the normal and superconducting states. The origin of the gap may be due to one of the two hole pockets around $\Gamma$ vanishes and the other one shrinks dramatically in the overdoped region according to ARPES experiments [18] and theories [23]. Under such a case, due to the lack of interband scattering between the hole and electron pockets, the imaginary part of the spin susceptibility is strongly suppressed and gives rise to the pseudogap behavior [46] which has been observed in nuclear magnetic resonance (NMR) [47] and NS [37] experiments. But in [46], the pseudogap is associated with the vanishing of one of three hole pockets around $\Gamma$, where experimentally there is only one hole pocket at this doping level as observed by ARPES [18]. The spin excitations in the superconducting state at $T = 0$ [Fig. 9.9(f)] are broader and weaker than those in the underdoped and optimally doped compounds, suggesting the importance of the hole pocket in enhancing the spin fluctuations.

At $x = 0.3$, both the two hole pockets around $\Gamma$ disappear [23], our calculations show that the SC order is completely suppressed and the spin fluctuations are extremely small [the inset of Fig. 9.9(e)]. This further indicates the correlation between the electronic band structure and magnetism, and supports the scenario that the spin fluctuations in the underdoped regime, which serve as a precursor to superconductivity, originate from quasiparticle scattering across the electron and hole pockets.

## 9.6 Vortex States

In 2010, an STM measurement on $Ba_{1-x}K_xFe_2As_2$ [48] has revealed, for the first time in the iron-pnictides, the existence of the Andreev bound states inside the vortex core with a systematic evolution: a single conductance peak appears at a negatively-biased voltage at the vortex center, which gradually evolves into two sub-peaks when moving away from the center, with a dominant spectral weight at negative bias. This negatively-biased conductance peak has not been observed in electron-doped $Ba(Fe_{1-x}Co_x)_2As_2$ [49] and is beyond current theoretical predictions where a peak appears at positive bias in a two-orbital model [21,50]. Therefore, it is important to develop a sound theory for the vortex states in the iron-pnictide superconductors. In view of this, we adopt a phenomenological model with competing SDW and $s_\pm$ superconductivity to study the vortex states in $Ba_{1-x}K_xFe_2As_2$ from the LDOS. The effect of SDW on the vortex states is also discussed, which we predict to be measurable by future experiments.

We use the same Hamiltonian as Eq. (9.2), except that $t_{i\mu j\nu}$ is replaced by $t'_{i\mu j\nu}$. In the presence of a magnetic field $B$ perpendicular to the plane, the hopping integral can be expressed as $t'_{i\mu j\nu} = t_{i\mu j\nu} \exp\left[i\frac{\pi}{\Phi_0}\int_j^i \boldsymbol{A}(\boldsymbol{r}) \cdot d\boldsymbol{r}\right]$, where $\Phi_0 = hc/2e$ is the superconducting flux quantum, and $\boldsymbol{A} = (-By, 0, 0)$ is the vector potential in the Landau gauge. The $s_\pm$ order parameter at site $j$ is $\Delta'_{j\nu} = (\Delta'_{j+\hat{x}+\hat{y}\nu j\nu} + \Delta'_{j-\hat{x}-\hat{y}\nu j\nu} + \Delta'_{j+\hat{x}-\hat{y}\nu j\nu} + \Delta'_{j-\hat{x}+\hat{y}\nu j\nu})/4$ where $\Delta'_{i\nu j\nu} = \Delta_{i\nu j\nu} \exp\left[i\frac{\pi}{\Phi_0}\int_i^{(i+j)/2} \boldsymbol{A}(\boldsymbol{r}) \cdot d\boldsymbol{r}\right]$. Magnetic unit cells are introduced where each unit cell accommodates two superconducting flux quantum and the linear dimension is $N_x \times N_y = 48 \times 24$, which is larger than the coherence length $\xi$ of the iron-pnictides [49]. In the following, we focus on two doping concentrations $x = 0.4$ and $0.3$, corresponding to the optimally doped and underdoped compounds, respectively.

At $x = 0.4$, first we choose $J_H = 0.2U$ such that, at $B = 0$, SDW is completely suppressed and $\Delta'_{j\nu}$ is homogeneous in real space. Figures 9.10(a) and 9.10(b) show the spatial variations of the reduced $s_\pm$ order parameter $\Delta^R_{j\nu} = |\Delta'_{j\nu}/\Delta'_{j\nu}(B = 0)|$ and the electron density $n_j = \sum_{\nu\sigma} n_{j\nu\sigma}$ plotted on a $24 \times 24$ lattice. The vortex center is located very close to site (11,12) and no SDW is induced. $\Delta^R_{j\nu}$ vanishes at the vortex center and starts to increase at the scale of the coherence length $\xi$ to its bulk value, but the increase is slower along the $\pi/4$ and $3\pi/4$ directions with respect to the underlying lattice. On the other hand, $n_j$ is strongly enhanced at the vortex center which is compensated by a depletion of electrons around two lattice spacings away from the center, after which $n_j$ decays also at the scale of $\xi$ to its bulk value, with no obviously slow variations along the $\pi/4$ and $3\pi/4$ directions. The zero-energy(ZE) LDOS plotted in Fig. 9.10(c) also peaks very close to site (11,12) and has the same fourfold rotational symmetry (RS) as $\Delta_{j\nu}^R$. In order to reveal the spatial variation of LDOS modulated by the vortex, in Fig. 9.10(d) we plot the LDOS at four typical positions along the black cut in Fig. 9.10(c). As we can see, at the vortex center, there is a remarkable negative-energy (NE) in-gap peak located at $-0.125\Delta$, which is precisely the same as observed in [48]. When moving away from the center, the peak will split into two in-gap peaks with a dominant spectral weight at negative energy. Finally, the LDOS evolves

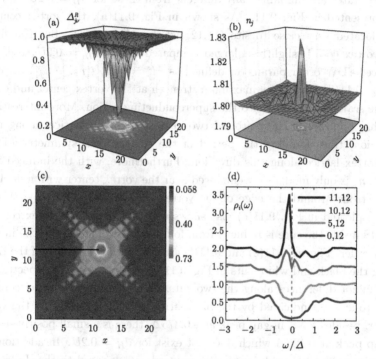

Fig. 9.10    From [51]. Spatial variations of (a) the reduced $s_\pm$ order parameter $\Delta_{j\nu}^R$, (b) electron density $n_j$, and (c) ZE LDOS map plotted on a $24 \times 24$ lattice. (d) The LDOS at four typical positions along the black cut in (c): at site (11,12) very close to the vortex center [the latter is in the center of the plaquette enclosed by sites (11,12), (12,12), (12,11) and (11,11)]; within the vortex core while away from the center (10,12); around the edge of a vortex (5,12); and far outside a vortex (0,12). The curves in (d) are displaced vertically for clarity and the gray dashed line indicates the position of zero energy. The doping concentration is $x = 0.4$ and $J_H = 0.2U$.

continuously into its bulk feature. The in-gap peak and evolution of the LDOS
clearly indicate the existence of the Andreev bound states inside the vortex
core, consistent with [48].

In order to study the effect of induced SDW on the vortex states, we
perform the calculation for $J_H = 0.23U$ since it has been shown that increasing
$J_H$ can lead to the formation of SDW [52]. Like the $J_H = 0.2U$ case, at $B = 0$,
SDW is completely suppressed and $\Delta'_{j\nu}$ is homogeneous in real space, but the
vortex states are fundamentally different from those for $J_H = 0.2U$ and they
are presented in Fig. 9.11. As shown in Fig. 9.11(a), the vortex center is
still located very close to site (11,12) where $\Delta^R_{j\nu}$ vanishes, but the size of
the vortex core is slightly enlarged compared to the $J_H = 0.2U$ case. The
induced SDW order parameter defined as $M^s_j = (-1)^{j_\nu}(n_{j\uparrow} - n_{j\downarrow})$ displayed
in Fig. 9.11(d) reaches its maximum strength at the vortex center and decays
at the scale of $\xi$ to zero into the superconducting region. More interestingly,
$M^s_j$ has opposite polarity around two nearest-neighbor vortices along the $\hat{x}$
direction, thus doubling the period of the translational symmetry (TS) of
the vortex lattice along this direction. Furthermore, with the induced SDW
order, $n_j$ is only moderately enhanced near the vortex center with a depletion
of electrons around the edge of the vortex core [see Fig. 9.11(b)]. The ZE
LDOS plotted in Fig. 9.11(c) also shows a slightly enlarged vortex core and
the RS inside the core is reduced from fourfold to twofold due to the induced
SDW order. Figures 9.11(e) and 9.11(f) are the spatial variations of the LDOS
along the black and white cuts in Fig. 9.11(c), respectively. The spectra have
only minor differences along the two cuts, mainly inside the vortex core and
close to $\omega = 0$ (indicated by the black arrow in Fig. 9.11(f)). At the vortex
center, besides a NE in-gap peak at $-0.375\Delta$, there is a small positive-energy
in-gap peak at $0.125\Delta$ which does not exist for $J_H = 0.2U$. In addition, the
intensity of the NE peak is strongly reduced as compared to the $J_H = 0.2U$
case and the spectral weight is transferred to form additional peaks outside
the gap as indicated by the red (or gray) arrows in Fig. 9.11(f). When moving
away from the center, the intensities of all these peaks decrease and finally the
LDOS evolves into its bulk feature. By comparing with the $J_H = 0.2U$ case,
we can identify that those two in-gap peaks are due to the Andreev bound
states while the others are due to the induced SDW order inside the core.

At $x = 0.3$, we choose $J_H = 0.32U$ so that, at $B = 0$, the $(\pi, 0)$ SDW coexists with the $s_\pm$ superconductivity. The vortex states are plotted in Fig. 9.12. $\Delta_{j\nu}^R$ shows a $\hat{y}$-axis oriented stripe-like feature with a modulation period of $8a$ [see Fig. 9.12(a)]. The size of the vortex core is further enlarged and elongated along the $\hat{y}$ direction. Moreover, the two vortex cores are dragged towards each other along the $\hat{x}$ direction with the vortex centers being located very close to sites $(15, 12)$ and $(32, 12)$, thus also doubling the period of the TS of the vortex lattice along this direction. $M_j^s$ shown in Fig. 9.12(d) behaves like nearly uniform stripes oscillating with a wavelength of $16a$. The vortex core is pinned at one of the ridges of SDW stripes where the SDW order is stronger than those at other sites. $n_j$ also exhibits a quasi-one-dimensional charge stripe behavior with a wavelength $8a$, exactly half that of the SDW along the $\hat{x}$ direction [see Fig. 9.12(b)]. The one-dimensional stripe-like modulations in $\Delta_{j\nu}^R$, $M_j^s$ and $n_j$ already exist at $B = 0$, which are quite similar to the cuprates except for a doubling of the period from $4a$ (for $\Delta_{j\nu}^R$ and $n_j$) and $8a$ (for $M_j^s$) in the cuprates [53] to $8a$ and $16a$ in the iron-pnictides. The origin of such stripes could be understood in terms of the existence of a nesting wave vector $q_A \sim 0.125\pi/a$ connecting the outer hole and electron pockets around the $\Gamma$ and $M$ points, respectively, along the $\hat{x}$ direction [see Fig. 9.13]. For proper values of $U$, $J_H$ and doping, this wave vector would modulate $M_j^s$ with SDW stripes along the $\hat{x}$ direction with period $2\pi/q_A = 16a$. The ZE LDOS in Fig. 9.12(c) also shows an enlarged vortex core, the doubling of the period of the TS of the vortex lattice along the $\hat{x}$ direction and the reduced RS from fourfold to twofold. Interestingly, although the two vortex cores are dragged towards each other, the ZE LDOS still peaks very close to sites $(11, 12)$ and $(36, 12)$, suggesting that even in the region where $\Delta_{j\nu}^R \neq 0$, there are ZE states contributing to the LDOS. The spatial variations of the LDOS plotted in Figs. 9.12(e) and 9.12(f) show that at the ZE LDOS peak position, there is a NE in-gap peak at $-0.375\Delta$, whose intensity is further reduced compared to that in Fig. 9.11(f) and the intensity decreases when moving away from the peak position, indicating that it is due to the Andreev bound states. There are also additional peaks outside the gap whose positions are similar to those marked by the red (or gray) arrows in Fig. 9.11(f). Their intensities vary drastically along the black cut in Fig. 9.12(c) while they barely change along the white cut, again suggesting that these peaks are due to the SDW order.

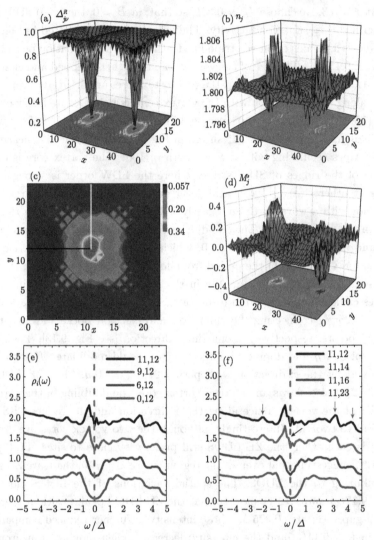

Fig. 9.11   From [51]. (a) and (b) are similar to Figs. 9.10(a) and 9.10(b), respectively, but are plotted on a $48 \times 24$ lattice. (c) is the same as Fig. 9.11(c). (d) The staggered magnetization $M_j^s$ plotted on a $48 \times 24$ lattice. (e) The LDOS at four typical positions along the black cut in (c). (f) is similar to (e), but is plotted along the white cut. The doping concentration is $x = 0.4$ and $J_H = 0.23U$.

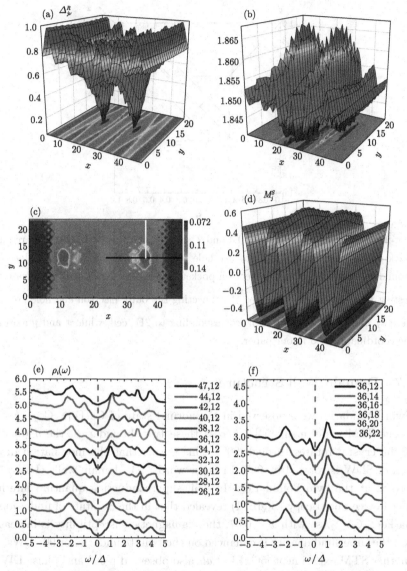

Fig. 9.12   From [51]. (a) and (b) are the same as Figs. 9.11(a) and 9.11(b), respectively. (c) is similar to Fig. 9.11(c), but is plotted on a $48 \times 24$ lattice. (d) is the same as Fig. 9.11(d). (e) The LDOS along the black cut in (c). (f) is similar to (e), but is plotted along the white cut. The doping concentration is $x = 0.3$ and $J_H = 0.32U$.

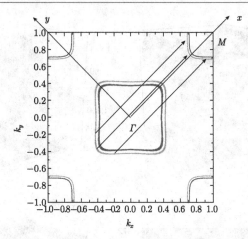

Fig. 9.13   Fermi surfaces at $x = 0.3$ and $T = 10^{-4}$, with $J_H = 0.32U$. The red and black pockets around the $\Gamma$ point are hole pockets, while the green and blue ones around the four $M$ points are electron pockets. The three black arrows indicate the nesting wave vector $\left(\pi + \dfrac{\pi}{8}, \pi + \dfrac{\pi}{8}\right)$ connecting the outer electron and hole pockets. Here $k_x$ and $k_y$ represent the BZ corresponding to 2Fe/cell, while $x$ and $y$ axes are the coordinates used in our paper.

## 9.7   Domain Wall Structure

Twin boundaries were observed in the normal state of $Ca(Fe_{1-x}Co_x)_2As_2$ in recent STM experiment [54]. Across these twin boundaries, the $a$ ($b$) axis of the crystal rotates through 90°. This means that the modulation direction of SDW is rotated by 90°. In other words, 90° domain walls (DWs) are formed at the twin boundaries. In [55], superconducting quantum interference device microscopy (SQIDM) revealed that in the SC state of underdoped $Ba(Fe_{1-x}Co_x)_2As_2$ with $x < 0.07$, the diamagnetic susceptibility is increased and the superfluid density is enhanced on the twin boundaries or 90° DWs. In another STM experiment [56], Li *et al.* also observed a 90° anti-phase DW in the parent compounds of iron pnictides, on which the LDOS is much higher than that in the interior of magnetic domains. Therefore, the existence of DWs in these underdoped 122 family of the iron-pnictide superconductors is universal and may affect strongly the electronic properties in the normal and

SC states. In order to interpret the complex DW structures observed by STM experiments on the iron-pnictides, we investigate the effect of strong Coulombic correlation on Fe sites, which leads to the formation of DW strutctures. By solving self-consistently the BdG equations, we obtain our results.

Our calculations show that 90° DWs exist in the large $U$ cases, with the values of electron density $n_i$, magnetic order $m_i$, and SC order parameter $\Delta_i$ being different on the DWs. We plot the phase diagram of DWs in the $U - J_H$ plane for the parent compound, i.e., $x = 0.0$. Figure 9.14(a) shows that, 90° DWs do not exist when $U \leqslant 4.7$ or $J_H \geqslant 2.8$. Figures 9.14(b) and 9.14(c) depict the fluctuation $\delta n = \sum_i |n_i - (2 + x)|/N$ of the electron density and the average magnetic moment $m = \sum_i |m_i|/N$ as functions of $U$ and $J_H$ in the presence of DW structure, respectively. When $U$ ($J_H$) increases [with $J_H$ ($U$) fixed], $\delta n$ becomes smaller while $m$ becomes larger. As $\delta n$ vanishes, DW structure also disappears.

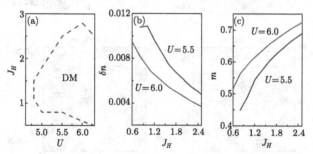

Fig. 9.14   From [57]. (a) The phase diagram in $U-J_H$ plane for zero doping. (b) The average value of $\delta n_i$ as a function of $U$ and $J_H$. (c) The average value of $|m_i|$ as a function of $U$ and $J_H$.

In the doped cases, taking $U = 4.8$, $J_H = 1.3$, and $V_{i\mu j\nu} = 1.1$ (for $\mu = \nu$ and $i$, $j$ being the NNN sites) as an example, in Fig. 9.15, we present the zero temperature images of electron density $n_i$, magnetic order $m_i$, and SC order $\Delta_i$ at electron dopings $x = 0.04$ and $0.16$ on a $N = 28 \times 28$ lattice with periodic boundary condition. From Figs. 9.15(b) and 9.15(e), we can see that there exist magnetic domain structures. Across the DWs, the modulation direction of magnetic order rotates through 90° at $x = 0.04$. At $x = 0.16$, the phase of magnetic order changes sign, and there are anti-phase DWs which have been predicted previously in [58] in the higher electron dopings. We observe that on both 90° DWs and anti-phase DWs, there are always higher electron densities $n_i$, as shown in Figs. 9.15(a) and 9.15(d). Therefore, it is expected that the superfluid density is enhanced on these DWs, which coincides with the observations of SQIDM experiments [55]. However, the SC order parameter $\Delta_i$ is larger on 90° DWs, while it is smaller on anti-phase DWs [see Figs. 9.15(c) and 9.15(f)]. Thus, except for the electron density, the magnetic and SC properties on the two kinds of DWs are very different.

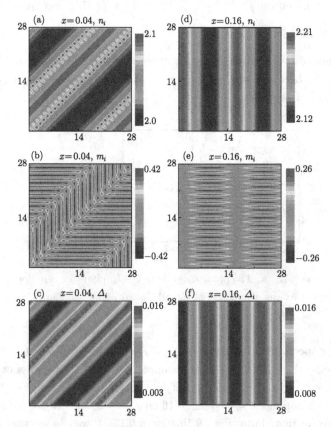

Fig. 9.15   From [57]. The images of electron density $n_i$, magnetic order $m_i$, and superconducting order $\Delta_i$ at electron dopings $x = 0.04$ and 0.16.

We would like to mention that when $x \leqslant 0.02$, $n_i$ and $m_i$ have similar patterns as those at $x = 0.04$, except for $\Delta_i = 0$. Therefore, the 90° DW structure also exists in the normal state of the FeAs-based superconductors, and this is consistent with the observations of STM experiments [54]. However, we do not get the solution of complex 90° anti-phase DW seen in the parent compounds [56], which cannot be formed under the periodic boundary condition.

Now we calculate the LDOS on and near the DWs in order to compare with the STM experiments. Fig. 9.16 shows the LDOS images on a 28 × 28 lattice at energies $\omega = \pm 0.02$ and electron dopings $x = 0.04, 0.16$ respectively. It is obvious that when $\omega = \pm 0.02$, $\rho_i(\omega)$ has the maximum value on the 90° DWs at $x = 0.04$, but has the minimum value on the anti-phase DWs at $x = 0.16$. Here we do not show the LDOS image at $x = 0.0$ since it is similar to that at $x = 0.04$.

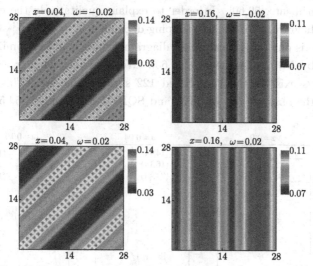

Fig. 9.16   From [57]. The images of LDOS $\rho_i(\omega)$ with electron dopings $x = 0.04$ and 0.16 at energies $\omega = \pm 0.02$.

However, the LDOS images vary with energy $\omega$. In Fig. 9.17, we show the energy dependence of the LDOS $\rho_i(\omega)$ at different sites along the line $y = 14$ at different electron dopings. At both $x = 0.0$ and 0.04, obviously, $\rho_i(\omega)$ at $(14, 14)$ on the 90° DW is always larger than that on the other sites when

$\omega \in (-0.18, 0.1)$. In contrast, $\rho_i(\omega)$ at $(1, 14)$ on the anti-phase DW is always smaller than that on the other sites when $\omega > -0.13$. The coherence peak at positive energy is higher than that at negative energy at both $x = 0.04$ and $0.16$ due to the coexistence of SDW and superconductivity [17,23]. The asymmetry of the coherence peaks was also observed by STM experiments [30].

When $x \leqslant 0.08$, the 90° domain walls always exist while the anti-phase magnetic domains show up at $x \geqslant 0.15$. In the range of $0.08 < x < 0.15$, the ground state is just like the small $U$ cases and the solution of anti-phase DW structure is a metastable state which has a slightly higher energy than that of the ground state. We have shown that strong electron correlations can induce robust doping-dependent magnetic structures in the normal and SC states of the FeAs-based superconductors. However, a large $U = 4.8$ leads to the disappearance of SDW order at a higher doping (0.21), compared to NMR experiments [59,60]. In order to explain qualitatively the experimental phase diagram, here we use a doping-dependent $U = 6.0 \exp\left[-(x/0.13)^2\right]$. Figure 9.18 is our theoretical phase diagram, which is indeed similar to that measured by NMR experiments [59,60]. We note that when $x \leqslant 0.06$, 90° DWs always exist in the FeAs-based 122 superconductors. This is consistent with the observations of STM and SQIDM experiments [54,55]. When

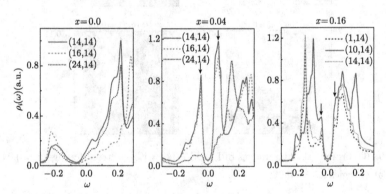

Fig. 9.17 From [57]. Energy dependence of LDOS $\rho_i(\omega)$ with electron dopings $x = 0.0$, 0.04, and 0.16 at the sites on and near DWs. The arrows point to the coherence peaks with maximum superconducting order parameter $\Delta_i$ in the panels of $x = 0.04$ and 0.16.

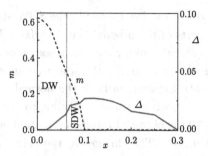

Fig. 9.18   From [57]. The average values of $|m_i|$, $\Delta_i$ as functions of doping $x$ in the case of doping-dependent $U = 6.0 \exp[-(x/0.13)^2]$.

$0.06 \leqslant x \leqslant 0.1$, the $2 \times 1$ SDW order and superconductivity coexist uniformly. For $0.1 \leqslant x \leqslant 0.3$, there is only pure SC state. Therefore, the anti-phase DWs predicted at higher dopings are not present in the Co-doped 122 superconductors. This is because at higher doping, the magnitude of $U$ becomes smaller, which is not able to sustain the state of the anti-phase DWs.

## 9.8   Conclusion

In conclusion, we have systematically studied the magnetic and SC properties in the 122 family of the iron-pnictide superconductors. Based on our effective four-band model, the derived SDW pattern, SC paring symmetry and phase diagram are all consistent with experimental observations. Especially, the FS evolution with doping calculated by our model agrees better with ARPES experiments than LDA.

As to the low-energy properties, first, we show that there exists asymmetry of the SC coherence peaks in the LDOS and this asymmetry varies with doping. In the underdoped region, the SC coherence peak at negative energy is lower than the one at positive energy. In the optimally doped sample, the intensities of the two peaks are almost equal while in the overdoped region, the situation is opposite to that in the underdoped case. The evolution of the asymmetry has been verified by recent STM experiment.

Then we investigated the doping dependence of spin excitations in $Ba(Fe_{1-x}Co_x)_2As_2$, ranging from the parent to overdoped regime. In the parent compound, the spin excitations are gapless in the paramagnetic state and become strongly suppressed at low energy in the SDW state due to the opening of gaps on most parts of the original Fermi surface. For underdoped and optimally doped samples, the spin gaps and resonances at $(\pi, \pi)$ only occur in the SC state. On the other hand, the spin excitations in the overdoped compound show gap-like behavior in both the normal and SC states due to the vanishing of one hole pocket around $\Gamma$, leading to a "pseudogap" behavior at this doping level. All the obtained results are in qualitative agreement with NS and NMR experiments. The changes in the spin dynamics at different doping levels may reflect changes in the electronic band structure and suggest a strong correlation between superconductivity and magnetism.

Furthermore, from the investigation of the vortex states in $Ba_{1-x}K_xFe_2As_2$ we found, in the optimally doped compound without induced SDW, there is a NE in-gap peak in the LDOS at the vortex center due to the Andreev bound states, which splits into two asymmetric in-gap peaks when moving away from the center. The effect of the induced SDW is mainly to reduce the intensity of the NE in-gap peak and transfer the spectral weight to form additional peaks outside the gap. For the underdoped sample where the SDW coexists with the $s_\pm$ superconductivity, the vortex cores are dragged towards each other along the $\hat{x}$ direction and the intensity of the NE in-gap peak is further reduced. The obtained result at $x = 0.4$ without induced SDW is in qualitative agreement with experiment and we propose future experiments on the near optimally doped and underdoped samples to verify the effect of the SDW on the vortex states. Furthermore, the asymmetry of the As atoms above and below the Fe layer is considered in our model, which is consistent with surface experiments such as ARPES and STM. Apparently this asymmetry is crucial in getting the NE in-gap peak to agree with experiment, since it has not been considered in other models. On the other hand, the disappearance of the Andreev bound states in electron-doped $Ba(Fe_{1-x}Co_x)_2As_2$ may be due to the induction of strong SDW order in the vortex states, which also needs to be verified by future experiments.

At last, we show that strong electron correlations can induce DWs, which separate the regions with different SDW orders. At zero or low electron doping, 90° DWs are robust against the on-site Coulombic interaction and Hund's coupling. On the DWs, there always exist larger electron densities. The results agree qualitatively with recent observations of STM and SQIDM in underdoped $(Ba,Ca)(Fe_{1-x}Co_x)_2As_2$. The existence of the DW structures leads to the nonuniformity of the electron density and the SC order parameter in real space. Therefore, the inhomogeneity of superconductivity is intrinsic in the underdoped iron pnictides. The origin of the DWs is due to the transition between two Fe sublattices rather than that between different orbitals. We note that all the folding models, where all the Fe sites are treated equivalently, do not possess such DW structures. The 90° DWs at low electron dopings and the electron densities on them agree well with the observations of STM and SQIDM experiments. We also predict that the anti-phase DWs should not be experimentally observable at any electron doping in the 122 superconductors.

## Acknowledgements

This work is supported by the Texas Center for Superconductivity at the University of Houston and the Robert A. Welch Foundation under Grant No. E-1146.

[1]  Y. Kamihara, T. Watanabe, M. Hirano and H. Hosono, J. Am. Chem. Soc. **130**, 3296 (2008).

[2]  Z. A. Ren, G. C. Che, X. L. Dong, J. Yang, W. Lu, W. Yi, X. L. Shen, Z. C. Li, L. L. Sun, F. Zhou and Z. X. Zhao, Europhys. Lett. **83**, 17002 (2008).

[3]  A. S. Sefat, R. Y. Jin, M. A. McGuire, B. C. Sales, D. J. Singh and D. Mandrus, Phys. Rev. Lett. **101**, 117004 (2008).

[4]  M. Rotter, M. Tegel and D. Johrendt, Phys. Rev. Lett. **101**, 107006 (2008).

[5]  M. Rotter, M. Tegel, D. Johrendt, I. Schellenberg, W. Hermes, and R. Pöttgen, Phys. Rev. B **78**, 020503(R) (2008).

[6]  I. I. Mazin, D. J. Singh, M. D. Johannes and M. H. Du, Phys. Rev. Lett. **101**, 057003 (2008).

[7]  D. K. Pratt, W. Tian, A. Kreyssig, J. L. Zarestky, S. Nandi, N. Ni, S. L. Bud'ko, P. C. Canfield, A. I. Goldman and R. J. McQueeney, Phys. Rev. Lett. **103**, 087001 (2009).

[8]  J. H. Chu, J. G. Analytis, C. Kucharczyk and I. R. Fisher, Phys. Rev. B **79**, 014506 (2009); F. Ning, K. Ahilan, T. Imai, A. S. Sefat, R. Y. Jin, M. A. McGuire, B. C. Sales and D. Mandrus, J. Phys. Soc. Jpn. **78**, 013711 (2009).

[9]  C. Lester, J. H. Chu, J. G. Analytis, S. C. Capelli, A. S. Erickson, C. L. Condron, M. F. Toney, I. R. Fisher and S. M. Hayden, Phys. Rev. B **79**, 144523 (2009).

[10]  H. Chen, Y. Ren, Y. Qiu, Wei Bao, R. H. Liu, G. Wu, T. Wu, Y. L. Xie, X. F. Wang, Q. Huang and X. H. Chen, Europhys. Lett. **85**, 17006 (2009).

[11] R. R. Urbano, E. L. Green, W. G. Moulton, A. P. Reyes, P. L. Kuhns, E. M. Bittar, C. Adriano, T. M. Garitezi, L. Bufaical and P. G. Pagliuso, Phys. Rev. Lett. **105**, 107001 (2010).

[12] Q. Huang, Y. Qiu, Wei Bao, M. A. Green, J.W. Lynn, Y. C. Gasparovic, T. Wu, G. Wu and X. H. Chen, Phys. Rev. Lett. **101**, 257003 (2008).

[13] Y. Nakai, T. Iye, S. Kitagawa, K. Ishida, S. Kasahara, T. Shibauchi, Y. Matsuda and T. Terashima, Phys. Rev. B **81**, 020503(R) (2010).

[14] H. Ding, P. Richard, K. Nakayama, K. Sugawara, T. Arakane, Y. Sekiba, A. Takayama, S. Souma, T. Sato, T. Takahashi, Z. Wang, X. Dai, Z. Fang, G. F. Chen, J. L. Luo and N. L. Wang, Europhys. Lett. **83**, 47001 (2008).

[15] K. Nakayama, T. Sato, P. Richard, Y.-M. Xu, Y. Sekiba, S. Souma, G. F. Chen, J. L. Luo, N. L. Wang, H. Ding and T. Takahashi, Europhys. Lett. **85**, 67002 (2009).

[16] K. Terashima, Y. Sekiba, J. H. Bowen, K. Nakayama, T. Kawahara, T. Sato, P. Richard, Y.-M. Xu, L. J. Li, G. H. Cao, Z.-A. Xu, H. Ding and T. Takahashi, Proc. Natl. Acad. Sci. U.S.A. **106**, 7330 (2009).

[17] D. G. Zhang, Phys. Rev. Lett. **103**, 186402 (2009).

[18] Y. Sekiba, T. Sato, K. Nakayama, K. Terashima, P. Richard, J. H. Bowen, H. Ding, Y. M. Xu, L. J. Li, G. H. Cao, Z. A. Xu and T. Takahashi, New J. Phys. **11**, 025020 (2009).

[19] M. M. Korshunov and I. Eremin, Phys. Rev. B **78**, 140509(R) (2008).

[20] A. M. Oles, G. Khaliullin, P. Horsch and L. F. Feiner, Phys. Rev. B **72**, 214431 (2005).

[21] H. M. Jiang, J. X. Li and Z. D. Wang, Phys. Rev. B **80**, 134505 (2009).

[22] J. X. Zhu, B. Friedman and C. S. Ting, Phys. Rev. B **59**, 3353 (1999).

[23] T. Zhou, D. G. Zhang and C. S. Ting, Phys. Rev. B **81**, 052506 (2010).

[24] P. Richard, K. Nakayama, T. Sato, M. Neupane, Y. M. Xu, J. H. Bowen, G. F. Chen, J. L. Luo, N. L. Wang, X. Dai, Z. Fang, H. Ding and T. Takahashi, Phys. Rev. Lett. **104**, 137001 (2010).

[25] Y. Ran, F. Wang, H. Zhai, A. Vishwanath and D. H. Lee, Phys. Rev. B **79**, 014505 (2009).

[26] T. Zhou, H. X. Huang, Y. Gao, J. X. Zhu and C. S. Ting, Phys. Rev. B **83**, 214502 (2011).

[27] D. Hsieh, Y. Xia, L. Wray, D. Qian, K. Gomes, A. Yazdani, G. F. Chen, J. L. Luo, N. L. Wang and M. Z. Hasan, arXiv: 0812.2289.

[28] K. K. Huynh, Y. Tanabe and K. Tanigaki, Phys. Rev. Lett. **106**, 217004 (2011).

[29] J. X. Zhu, W. Kim, C. S. Ting and J. P. Carbotte, Phys. Rev. Lett. **87**, 197001(2001); W. Kim, J. X. Zhu, J. P. Carbotte and C. S. Ting, Phys. Rev. B **65**, 064502 (2002).

[30]  S. H. Pan *et al.*, private communication.

[31]  A. D. Christianson, M. D. Lumsden, S. E. Nagler, G. J. MacDougall, M. A. McGuire, A. S. Sefat, R. Jin, B. C. Sales and D. Mandrus, Phys. Rev. Lett. **103**, 087002 (2009).

[32]  R. A. Ewings, T. G. Perring, R. I. Bewley, T. Guidi, M. J. Pitcher, D. R. Parker, S. J. Clarke and A. T. Boothroyd, Phys. Rev. B **78**, 220501(R) (2008).

[33]  K. Matan, R. Morinaga, K. Iida and T. J. Sato, Phys. Rev. B **79**, 054526 (2009).

[34]  D. Parshall, K. A. Lokshin, J. Niedziela, A. D. Christianson, M. D. Lumsden, H. A. Mook, S. E. Nagler, M. A. McGuire, M. B. Stone, D. L. Abernathy, A. S. Sefat, B. C. Sales, D. G. Mandrus and T. Egami, Phys. Rev. B **80**, 012502 (2009).

[35]  M. D. Lumsden, A. D. Christianson, D. Parshall, M. B. Stone, S. E. Nagler, G. J. MacDougall, H. A. Mook, K. Lokshin, T. Egami, D. L. Abernathy, E. A. Goremychkin, R. Osborn, M. A. McGuire, A. S. Sefat, R. Jin, B. C. Sales and D. Mandrus, Phys. Rev. Lett. **102**, 107005 (2009).

[36]  D. S. Inosov, J. T. Park, P. Bourges, D. L. Sun, Y. Sidis, A. Schneidewind, K. Hradil, D. Haug, C. T. Lin, B. Keimer and V. Hinkov, Nature Physics **6**, 178 (2010).

[37]  K. Matan, S. Ibuka, R. Morinaga, S. X. Chi, J. W. Lynn, A. D. Christianson, M. D. Lumsden and T. J. Sato, Phys. Rev. B **82**, 054515 (2010).

[38]  C. Lester, J. H. Chu, J. G. Analytis, T. G. Perring, I. R. Fisher and S. M. Hayden, Phys. Rev. B **81**, 064505 (2010).

[39]  W. C. Lv, F. Krüger and P. Phillips, Phys. Rev. B **82**, 045125 (2010).

[40]  M. A. Kastner and R. J. Birgeneau, Rev. Mod. Phys. **70**, 897 (1998).

[41]  T. A. Maier and D. J. Scalapino, Phys. Rev. B **78**, 020514(R) (2008).

[42]  J. Knolle, I. Eremin, A. V. Chubukov and R. Moessner, Phys. Rev. B **81**, 140506(R) (2010).

[43]  M. M. Korshunov and I. Eremin, Europhys. Lett. **83**, 67003 (2008).

[44]  Y. Gao, T. Zhou, C. S. Ting and W. P. Su, Phys. Rev. B **82**, 104520 (2010).

[45]  R. J. McQueeney, S. O. Diallo, V. P. Antropov, G. D. Samolyuk, C. Broholm, N. Ni, S. Nandi, M. Yethiraj, J. L. Zarestky, J. J. Pulikkotil, A. Kreyssig, M. D. Lumsden, B. N. Harmon, P. C. Canfield and A. I. Goldman, Phys. Rev. Lett. **101**, 227205 (2008); S. O. Diallo, V. P. Antropov, T. G. Perring, C. Broholm, J. J. Pulikkotil, N. Ni, S. L. Bud'ko, P. C. Canfield, A. Kreyssig, A. I. Goldman and R. J. McQueeney, Phys. Rev. Lett. **102**, 187206 (2009).

[46]  H. Ikeda, R. Arita, and J. Kuneš, Phys. Rev. B **81**, 054502 (2010), Phys. Rev. B **82**, 024508 (2010).

[47] F. L. Ning, K. Ahilan, T. Imai, A. S. Sefat, M. A. McGuire, B. C. Sales, D. Mandrus, P. Cheng, B. Shen and H. H. Wen, Phys. Rev. Lett. **104**, 037001 (2010).

[48] L. Shan, Y. L. Wang, B. Shen, B. Zeng, Y. Huang, A. Li, D.Wang, H. Yang, C. Ren, Q. H. Wang, S. H. Pan and H. H. Wen, Nature Physics **7**, 325 (2011).

[49] Y. Yin, M. Zech, T. L. Williams, X. F. Wang, G. Wu, X. H. Chen and J. E. Hoffman, Phys. Rev. Lett. **102**, 097002 (2009).

[50] X. Hu, C. S. Ting and J. X. Zhu, Phys. Rev. B **80**, 014523 (2009); M. A. N. Araújo, M. Cardoso and P. D. Sacramento, New J. Phys. **11**, 113008 (2009).

[51] Y. Gao, H. X. Huang, C. Chen, C. S. Ting and W. P. Su, Phys. Rev. Lett. **106**, 027004 (2011).

[52] S. Zhou and Z. Q. Wang, Phys. Rev. Lett. **105**, 096401 (2010).

[53] H. Y. Chen and C. S. Ting, Phys. Rev. B **68**, 212502 (2003); W. D. Wise, M. C. Boyer, K. Chatterjee, T. Kondo, T. Takeuchi, H. Ikuta, Y. Y. Wang, E. W. Hudson, Nature Physics **4**, 696 (2008).

[54] T. M. Chuang, M. P. Allan, J. Lee, Y. X., N. Ni, S. L. Budko, G. S. Boebinger, P. C. Canfield and J. C. Davis, Science **327**, 181 (2010).

[55] B. Kalisky, J. R. Kirtley, J. G. Analytis, J. H. Chu, A. Vailionis, I. R. Fisher and K. A. Moler , Phys. Rev. B **81**, 184513 (2010).

[56] G. R. Li, X. B. He, A. Li, S. H. Pan, J. D. Zhang, R. Y. Jin, A. S. Sefat, M. A. McGuire, D. G. Mandrus, B. C. Sales and E. W. Plummer, arXiv:1006.5907.

[57] H. X. Huang, D. G. Zhang, T. Zhou and C. S. Ting, Phys. Rev. B **83**, 134517 (2011).

[58] I. I. Mazin and M. D. Johannes, Nature Phys. **5**, 141 (2009).

[59] Y. Laplace, J. Bobroff, F. Rullier-Albenque, D. Colson and A. Forget, Phys. Rev. B **80**, 140501 (2009).

[60] M. H. Julien, H. Mayaffre, M. Horvatic, C. Berthier, X. D. Zhang, W. Wu, G. F. Chen, N. L. Wang and J. L. Luo, Europhys. Lett. **87**, 37001 (2009).

# 10

# Superconductivity at 41 K and its Competition with Spin-Density-Wave Instability in Layered CeO$_{1-x}$F$_x$FeAs*

G. F. Chen, Z. Li, D. Wu, G. Li, W. Z. Hu, J. Dong, P. Zheng,
J. L. Luo and N. L. Wang

*Beijing National Laboratory for Condensed Matter Physics, Institute of Physics,
Chinese Academy of Sciences,
Beijing 100080, People's Republic of China*

A series of layered CeO$_{1-x}$F$_x$FeAs compounds with $x = 0$ to 0.20 are synthesized by the solid state reaction method. Similar to the LaOFeAs, the pure CeOFeAs shows a strong resistivity anomaly near 145 K, which was ascribed to the spin-density-wave instability. $F$ doping suppresses this instability and leads to the superconducting ground state. Most surprisingly, the superconducting transition temperature could reach as high as 41 K. Such a high $T_c$ strongly challenges the classic BCS theory based on the electron-phonon interaction. The closeness of the superconducting phase to the spin-density-wave instability suggests that the magnetic fluctuation plays a key role in the superconducting pairing mechanism. The study also reveals that the Ce $4f$ electrons form local moments and are ordered antiferromagnetically below 4 K, which could coexist with superconductivity.

The recent discovery of superconductivity with a transition temperature of 26K in LaO$_{1-x}$F$_x$FeAs system [1] has generated tremendous interest in the scientific community. Besides a relatively high transition temperature, the system displays many interesting properties. Among others, the presence of

---

competing ordered ground states is one of the most interesting phenomena [2]. The pure LaOFeAs itself is not superconducting but shows an anomaly near 150 K in both resistivity and dc magnetic susceptibility [1]. This anomaly was shown to be caused by the spin-density-wave (SDW) instability [2–4]. Electron doping by $F$ suppresses the SDW instability and yields the superconductivity. Here we show that similar competing orders exist in another rare-earth transition metal oxypnictide $Ce(O_{1-x}F_x)FeAs$. Most surprisingly, the superconducting transition temperature in this system could reach as high as 41 K. Except for cuprate superconductors, $T_c$ in such iron-based compounds has already become the highest.

The rather high superconducting transition temperature has several important implications. First, the electronphonon interaction needs to be carefully examined. There have been several first-principle calculations indicating that the electron-phonon coupling is not strong enough to explain the superconductivity in $LaO_{1-x}F_xFeAs$ [5–7]. However, an opposite opinion also exists [8]. The substantially higher $T_c$ here poses a more serious challenge to the classic phonon-mediated pairing mechanism. Second, the rare-earth Ce-based compounds usually show hybridization between the localized $f$ electrons and itinerant electrons. This often leads to a strong enhancement of carrier effective mass at low temperature. Even for $4d$ transition metal oxypnictide with the same type of structure as CeOFeAs, a recent report indicates that the electronic specific heat coefficient of Ce-based CeORuP ($\gamma = 77$ mJ/mol K$^2$) is 20 times higher than the value of La-based LaORuP ($\gamma = 3.9$ mJ/mol K$^2$) [9]. The hybridization tends to cause various ordered states at low temperature, like ferromagnetic (FM) or antiferromagnetic (AFM) ordering. Although the superconducting state could occur in Ce-based materials, the superconducting transition temperature is usually very low. The highest superconducting transition temperature is only 6.1 K achieved in $CeRu_2$ [10]. The very high superconducting transition temperature obtained here on $Ce(O_{1-x}F_x)FeAs$ offers an opportunity to examine the role played by Ce $4f$ electrons. Third, the superconducting phase is very close to the spin-density-wave instability. This indicates that the magnetic fluctuation plays a key role in the superconducting pairing mechanism.

A series of layered $CeO_{1-x}F_xFeAs$ compounds with $x = 0$, 0.04, 0.08, 0.12, 0.16, and 0.20 are synthesized by the solid state reaction method using CeAs,

Fe, $CeO_2$, $CeF_3$, $Fe_2As$ as starting materials. CeAs was obtained by reacting
Ce chips and As pieces at 500°C for 15 h and then 850°C for 5 h. The raw
materials were thoroughly mixed and pressed into pellets. The pellets were
wrapped with Ta foil and sealed in an evacuate quartz tube. They were then
annealed at 1150°C for 50 h. The resulting samples were characterized by a
powder X-ray diffraction (XRD) method with Cu $K\alpha$ radiation at room temperature. The XRD patterns for the parent $(x = 0)$ and $x = 0.16$ compounds
are shown in Fig. 10.1, which could be well indexed on the basis of tetragonal ZrCuSiAs-type structure with the space group $P4/nmm$. A tiny extra
(impurity) peak at $2\theta \approx 27°$ is observed only in $F$-doped sample. The lattice
parameters for the parent and $x = 0.16$ compounds obtained by a least-square
fit to the experimental data are $a = 3.996$ Å, $c = 8.648$ Å, and $a = 3.989$ Å,
$c = 8.631$ Å, respectively. Compared to the undoped phase CeOFeAs, the
apparent reduction of the lattice volume upon $F$ doping indicates a successful
chemical substitution. A careful inspection of the two XRD patterns in Fig.
10.1 also reveals some differences in the peak intensities for the two samples.
As we found that fluorine doping reduces the reaction temperature, the grain
sizes are expected to be larger in the $F$-doped samples. This may explain the
enhanced intensities of refractions from certain crystalline planes.

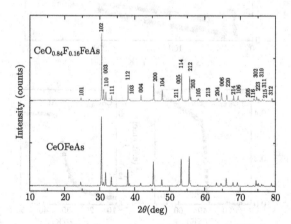

Fig. 10.1   X-ray powder diffraction patterns of the pure CeOFeAs and $CeO_{0.84}F_{0.16}$
FeAs compounds.

Standard 4-probe dc resistivity and ac susceptibility measurements were preformed down to 1.8 K in a Physical Property Measurement System (PPMS) of Quantum Design company. Figure 10.2(a) shows the temperature dependence of the resistivity. The pure CeOFeAs sample has rather high dc resistivity value. The resistivity increases slightly with decreasing temperature, but below roughly 145 K, the resistivity drops steeply. After $F$ doping, the overall resistivity decreases and the 145 K anomaly shifts to lower temperature and becomes less pronounced. At higher $F$ doping, the anomaly disappears and a superconducting transition occurs. The highest $T_c$ (41 K) is obtained at $x = 0.16$, which can be seen clearly from an expanded plot of the temperature-dependent resistivity curve [Fig. 10.2(b)]. $T_c$ drops slightly with further $F$

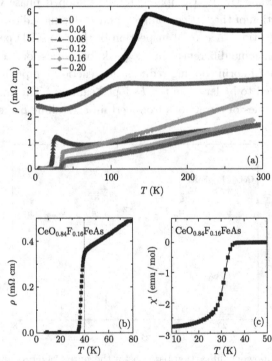

Fig. 10.2   (a) The electrical resistivity vs temperature for a series of CeO$_{1-x}$F$_x$FeAs. (b) $T$-dependent resistivity in an expanded region for $x = 0.16$ sample. The superconducting transition with sharp onset temperature at 41 K is seen. (c) Real part of $T$-dependent ac magnetic susceptibility.

doping. The bulk superconductivity in $F$-doped CeOFeAs is confirmed by dc magnetic susceptibility measurements. Figure 10.2(c) shows the real part $\chi'$ of ac susceptibility in a temperature range near $T_c$ for the $x = 0.16$ sample. The superconducting volume fraction is estimated to be about 60%. Figure 10.3(a) is the phase diagram showing the resistivity anomaly, i.e., SDW transition (circle) and superconducting transition (square) temperatures as a function of $F$ content.

An important parameter to characterize superconductivity is the upper critical field $H_{c2}(0)$. In our earlier study on $LaO_{0.9}F_{0.1-\delta}FeAs$ superconductor with an onset $T_c = 26$ K, we already found a rather high upper critical field $H_{c2}(0)$ over 50 T [11]. Here we would expect much higher $H_{c2}(0)$ in Ce-based compounds owing to their substantially higher $T_c$. We measured the temperature-dependent resistivity of a $x = 0.12$ sample with $T_c$ onset close to 40 K under different magnetic fields. As shown in Fig. 10.3(b), $T_c$ was suppressed only by several Kelvins at 14 T (which is the highest magnetic field available in our PPMS system). A precise determination of $H_{c2}(0)$ apparently requires measurement under much higher field [11].

The resistivity behavior of the pure CeOFeAs is very similar to that of LaOFeAs, except that a resistivity upturn was observed in the later compound at low temperature. As we demonstrated earlier, the anomaly at 150 K is caused by spin-density-wave instability, and a gap opens below the transition temperature due to the Fermi surface nesting [2]. To confirm the same origin for the anomaly, we performed infrared measurement on Bruker 66v/s spectrometer in the frequency range from 40 cm$^{-1}$ to 15000 cm$^{-1}$ at different temperatures, and derived the conductivity from Kramers-Kronig transformations. Figure 10.4 shows the reflectance and conductivity spectra in a far-infrared region. As expected, CeOFeAs shares very similar optical response behavior as LaOFeAs. Most notably, the reflectance below 400 cm$^{-1}$ is strongly suppressed at low frequency below the phase transition temperature, which is a strong indication for the formation of an energy gap. However, the low-frequency reflectance still increases fast towards unity at zero frequency, indicating a metallic behavior even below the phase transition, being consistent with the dc resistivity measurement which reveals an enhanced conductivity. The data indicate clearly that the Fermi surfaces are only partially gapped.

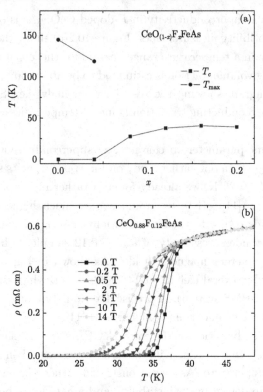

Fig. 10.3    (a) The phase diagram showing the anomaly (circle) and superconducting transition (square) temperatures as a function of $F$ content. (b) The resistivity vs temperature curves under selected magnetic fields.

We notice that, among different reported superconducting systems in such layered transition metal oxypnictides, the $LaO_{1-x}F_xFeAs$ and $CeO_{1-x}F_xFeAs$ systems share remarkably similar character: the presence of competing ground states. When the SDW order is destroyed by electron doping, superconductivity could occur at a much higher temperature. This gives a hint on how to search for materials with potentially higher $T_c$. The interplay between superconductivity and spin-density-wave instability is thus of central interest in those systems.

To get insight into whether the rare-earth element Ce $4f$ electrons hybridize with the itinerant Fe $3d$ electrons at low temperature, we measured the

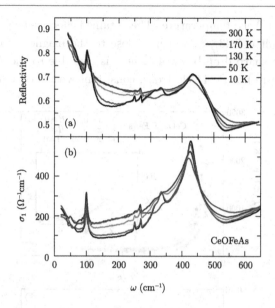

Fig. 10.4  The reflectance (a) and conductivity (b) spectra in the far-infrared region at different temperatures for the pure CeOFeAs sample.

low-$T$ specific heat. To our surprise, another magnetic ordering was revealed in those Ce-based samples. Figure 10.5(a) shows the plot of $C/T$ as a function of temperature for pure CeOFeAs at $H = 0$ and 5 T, and 16% $F$-doped sample ($CeO_{0.84}F_{0.16}FeAs$) at $H = 0$, respectively. For nonsuperconducting CeOFeAs, a sharp $\lambda$-shape peak at 3.7 K is observed under zero magnetic field. The peak shifts to 2.8 K under a magnetic field of 5 T. This clearly indicates that a long-range antiferromagnetic ordering occurs at low temperature. There is a very weak effect in dc resistivity at the AFM transition temperature. For LaOFeAs without the rare-earth $4f$ electrons, there is no such specific heat anomaly at low temperature [2], indicating unambiguously that the AFM transition for CeOFeAs is originated from the ordering of Ce $4f$ moments. No significant enhancement of electronic specific coefficient is observed from the high $T$ specific heat. We also measured the dc magnetic susceptibility of the parent CeOFeAs sample in the PPMS system; the data are shown in Fig. 10.5(b). By fitting the data with the Curie-Weiss law, $\chi(T) = \chi_0 + C/(T + \theta)$, where $\chi_0$ is a temperature independent susceptibility, $C$ the Curie constant, and $\theta$ the Curie-Weiss temperature. The obtained effective magnetic moments per formula unit is 2.43 $\mu_B$, which is close to the magnetic moment of free Ce3+ion. The sharp drop below 4 K here is also due to the AFM ordering (Néel temperature) of Ce $4f$ electrons, consistent with specific heat data.

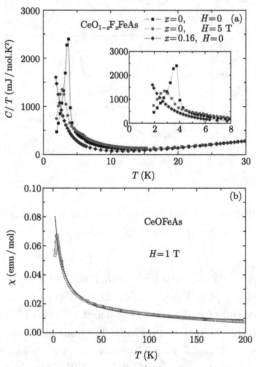

Fig. 10.5 (a) The plot of $C/T$ vs $T$ for the pure CeOFeAs sample at $H = 0$ and 5 T, and the CeO$_{0.84}$F$_{0.16}$FeAs sample at $H = 0$, respectively. The inset shows the plot in an expanded low temperature range. (b) The $T$-dependent magnetic susceptibility of pure CeOFeAs sample. The red (or gray) curve is a fit to the Curie-Weiss law. The sharp drop below 4 K is due to AFM ordering.

For the 16% $F$ doped superconducting sample, we also observe the onset signature of the AFM ordering down to 1.8 K, the lowest measured temperature. Then, the AFM transition must occur below this temperature. The data strongly suggest that the high-temperature superconductivity coexists with the AFM ordering of Ce $4f$ local moments. This coexistence implies that the exchange interaction between the Ce $4f$ moments and the itinerant Fe $3d$ electrons is very weak. So, there is no appreciable mixing or hybridization between them in the present systems.

To summarize, we have synthesized a series of rare-earth based transition metal oxypnictide $CeO_{1-x}F_xFeAs$ compounds. The superconducting transition temperature could be as high as 41 K. This very high superconducting transition temperature strongly challenges the pairing mechanism based on the electron-phonon interaction. Similar to the LaOFeAs, the pure CeOFeAs shows a strong resistivity anomaly near 145 K, which was ascribed to the spin-density-wave instability. $F$ doping suppresses this instability and leads to the superconducting ground state with rather high $T_c$. The very interesting interplay between the superconducting phase and the spin-density-wave instability strongly suggests that the magnetic fluctuation plays a key role in the superconducting pairing mechanism. Furthermore, the study reveals that the Ce $4f$ electrons form local moments and ordered antiferromagnetically below 4 K which could coexist with superconductivity.

We acknowledge the support from Y. P. Wang and Li Lu, and valuable discussions with Z. Fang and T. Xiang. This work is supported by the National Science Foundation of China, the Knowledge Innovation Project of the Chinese Academy of Sciences, and the No. 973 Project of the Ministry of Science and Technology of China.

*Note added.* —After we completed this work, we learned of an independent work on another rare-earth Sm-based superconductor $SmFeAsO_{1-x}F_x(x = 0.1)$ by Chen *et al.* [12].

[1] Y. Kamihara, T. Watanabe, M. Hirano and H. Hosono, J. Am. Chem. Soc. **130**, 3296 (2008).

[2] J. Dong, H. J. Zhang, G. Xu, Z. Li, G. Li, W. Z. Hu, D. Wu, G. F. Chen, X. Dai, J. L. Luo, Z. Fang and N. L. Wang, Europhysics Letters **83**, 27006 (2008).

[3] C. de la Cruz, Q. Huang, J. W. Lynn, J. Y. Li, W. Ratcliff, J. L. Zarestky, H. A. Mook, G. F. Chen, J. L. Luo, N. L. Wang and P. C. Dai, Nature **453**, 899 (2008).

[4] M. A. McGuire, A. D. Christianson, A. S. Sefat, R. Jin, E. A. Payzant, B. C. Sales, M. D. Lumsden and D. Mandrus, arXiv:cond-mat/ 0804.0796.

[5]   K. Haule, J. H. Shim and G. Kotliar, Phys. Rev. Lett. **100**, 226402 (2008).

[6]   L. Boeri, O. V. Dolgov and A. A. Golubov, Phys. Rev. Lett. **101**, 026403 (2008).

[7]   I. I. Mazin, D. J. Singh, M. D. Johannes and M. H. Du, Phys. Rev. Lett. **101**, 057003 (2008).

[8]   H. Eschrig, arXiv:cond-mat/0804.0186.

[9]   C. Krellner, N. S. Kini, E. M. Brüning, K. Koch, H. Rosner, M. Nicklas, M. Baenitz and C. Geibel, Phys. Rev. B **76**, 104418 (2007).

[10]   T. Matthias, H. Suhl and E. Corenzwit, Phys. Rev. Lett. **1**, 449 (1958).

[11]   G. F. Chen, Z. Li, G. Li, J. Zhou, D. Wu, J. Dong, W. Z. Hu, P. Zheng, Z. J. Chen, J. L. Luo and N. L. Wang, Phys. Rev. Lett. **101**, 057007 (2008).

[12]   X. H. Chen *et al.*, Nature **453**, 761 (2008).

# 11
# Effect of a Zn Impurity and the Superconducting Pairing Symmetry in Iron-Based Superconductors

Y. K. Li, X. Lin, Q. Tao, G. H. Cao and Z. A. Xu

*Department of Physics, Zhejiang University, Hangzhou 310027, People's Republic of China*

The effect of a non-magnetic Zn impurity on superconductivity may be completely different in the superconductors with different pairing symmetry. Thus the doping effect of the non-magnetic element Zn becomes an important supplementary means to explore the iron-based superconductors. We found that the non-magnetic Zn impurities severely suppresses the spin density wave (SDW) in the parent compounds. However, in the optimal doped $LaFeAsO_{0.9}F_{0.1}$ superconductors ($T_c = 26$ K), slight Zn doping has little effect on the superconducting (SC) transition temperature ($T_c$). Further studies showed that the doping effect of the impurity Zn is closely related with the concentration of F doping in the $LaFe_{1-x}Zn_xAsO_{1-y}F_y$ system. In the presence of a Zn impurity, $T_c$ increases in the under-doped regime, remains unchanged in the optimally doped regime and is severely suppressed in the over-doped regime. Comparing with the theoretical models, we propose that the superconducting pairing symmetry may change with the carrier concentration (F content) in the system. In the under-doped and optimally-doped regimes, we believe that the SC pairing may be the usual $s++$ pairing, where the non-magnetic impurities have little effect on $T_c$ according to the Anderson's theorem. In the under-doped regime, $T_c$ increases due to the suppression of the SDW order by Zn impurity where SC order may compete with the SDW order. In the over-doped regime, the SC gap should have nodes, and the SC pairing symmetry may be $d$−wave or $s+/-$ wave pairing. These observations help to understand the pairing mechanism of iron-based superconductors.

## 11.1   Introduction

Since the discovery of superconductivity at 26 K in $LaFeAsO_{0.9}F_{0.1}$ in 2008
[1], several iron-based superconducting systems containing FeAs layers have
been found [2−9] and $T_c$ has been raised up to 56 K[2−6]. FeSe-based superconductors with the similar layered structures have also been found [10−12].
The pairing symmetry of iron-based superconductors is one of the essential
issues to understand the superconducting (SC) mechanism. The SC gaps of
the 1111 phase and 122 phase iron-based superconductors were found to be
isotropic by angle-resolved photoemission spectroscopy (ARPES)[13−14], but
some other experiments, such as penetration depth [15−16] and nuclear magnetic resonance (NMR) [17−18], found evidences for nodal SC gaps. Up to
now the pairing symmetry of iron-based superconductors is still an open question. Meanwhile, many theoretical studies have proposed that the multi-band
characteristics of iron-based superconductors may cause the sign change of
SC order parameters in different SC bands, i.e. so-called $s+/-$ wave pairing, and the coupling of $s+$ and $s-$ SC bands plays a very crucial role in
superconductivity [19−22].

   According to the Anderson's theorem [23], non-magnetic impurities do not
cause a severe pair-breaking effect in a conventional $s$-wave superconductor,
and only the magnetic impurities cause the significant pair-breaking. However, the later experiments found that Zn impurity as a non-magnetic impurity
doped into the $CuO_2$ planes would cause remarkable pairing-breaking effect in
the $d$-wave pairing SC cuprates [24−25]. Some theoretical works proposed that
non-magnetic impurities in $d$-wave pairing superconductors would have same
pairing-breaking effect as magnetic impurities in $s$-wave pairing superconductors [26]. For the newly discovered iron-based superconductors, theoretical
analysis also points out that in the $s+/-$ wave SC state, non-magnetic impurities can severely break the pairing similar to the effect in high-$T_c$ cuprates
with $d$-wave pairing [21−22]. The zinc ion has a stable $3d^{10}$ configuration
in the alloy, without local magnetic moment, and can serve as the suitable
non-magnetic impurity in the superconductors for this study. In addition, Zn
doping does not affect the carrier concentration of the system, as found in
the Zn-doped high-temperature cuprates, thus is an ideal non-magnetic impurity [24−25]. The first-principles calculations also pointed out [27], in the
LaFeAsO structure, $3d$ electron orbits of Zn are far below Fermi energy, in-

dicating its $3d$ electrons are completely local, and will not make contribution to charge carriers. Therefore, Zn doping in the iron-based superconductors can also be taken as a non-magnetic impurity, which will not affect the carrier concentration in the system, and the doping effect can help to determine the SC pairing symmetry.

Our group first reported that Zn impurities have remarkable effect on the SDW order in the LaFeAsO parent compound and the Zn impurity effect on $T_c$ in optimal doping LaFeAsO$_{0.9}$F$_{0.1}$ superconductors [28]. Further study on the Zn impurity effect for different electron-doped LaFeAs(O,F) superconductors indicated that for different F doping concentration, the Zn-impurity has different effect on superconductivity [29]. In this article we will review the Zn-impurity effect in the iron-based superconductors and its implication to the pairing symmetry based on the theoretical models. We propose that the SC pairing symmetry in iron-based superconductors may change with the carrier concentration in the system.

## 11.2  Experimental

Polycrystalline samples of LaFe$_{1-x}$Zn$_x$AsO$_{1-y}$F$_y$ were synthesized by a two-step solid-state reaction method. As a first step, the intermediate materials of LaAs, FeAs, and Fe$_2$As were synthesized in vacuum. LaAs was presynthesized by reacting stoichiometric La and As powders after mixing in evacuated quartz tubes, calcinating at 873 K for 10 h, then sintering at 1173 K for 15 h. Using the same method, FeAs and Fe$_2$As were prepared by reacting stoichiometric Fe powders and As powders at 873 K and 1073 K for 10 h, respectively. In the second step, the final samples were synthesized from the intermediate materials. La$_2$O$_3$ was baked at 1193 K for 10 h to remove the water of crystallization, then the powders of these intermediate materials (LaAs, La$_2$O$_3$, FeAs, Fe$_2$As, ZnO, LaF$_3$) were weighed according to the stoichiometric ratio and thoroughly mixed in an agate mortar. The mixtures were pressed into pellets under a pressure of 2000 kg/cm$^2$. All the processes were operated in a glove box filled with high purity argon to ensure that the sample was not oxidized. The pellets were sealed in evacuated quartz tubes at $10^{-3}$ Pa and heated uniformly at 1433 K for 48 h, and then naturally cooled to room temperature. The final samples were black pellets.

We used powder X-ray diffraction (XRD) to determine the structure and phase purity of samples, and used Rietveld refinement fitting to obtain the lattice parameters. The electrical resistivity was measured by a standard four-terminal method. The temperature dependence of dc magnetic susceptibility was measured on a Quantum Design magnetic property measurement system (MPMS-5). Thermopower was measured by using a steady-state technique, and the temperature difference at both ends was obtained by the differential thermocouple. Hall effect was measured on a Quantum Design Physical Property Measurement System (PPMS-9).

## 11.3    Results and Discussion

### 11.3.1    The Suppression Effect of Zn Impurity on SDW Order in LaFeAsO Parent Compound

We have synthesized a series of Zn-doped polycrystalline samples of $LaFe_{1-x}Zn_xAsO$ and analyzed the structure of the samples by powder X-ray diffraction (XRD). According to the room temperature XRD pattern, there is a minor LaZnAsO impure phase, and a phase separation of LaZnAsO from the uniform main phase could occur for large Zn content ($x > 0.05$), indicating that Zn solid solubility in this system cannot exceed 10% [28]. Only for the slightly Zn-doped compounds ($x < 0.1$), the samples are essentially single phase. The lattice parameters are dependent on the Zn content, i.e., the $a$-axis increases slightly with increasing $x$, and the $c$-axis shrinks slightly, which indicates that Zn has successfully been doped into the crystal lattice. For a larger Zn content ($x > 0.1$), the change in the lattice constants $a$ and $c$ becomes saturated, which is in agreement with the phase separation.

The temperature dependence of resistivity were measured for the $LaFe_{1-x}Zn_xAsO$ samples. As shown in Fig. 11.1(a), the resistivity of the undoped parent compound shows an anomaly caused by SDW phase transition and structural phase transition, and it shows semiconductor-like behavior at low temperatures. Even with addition of only 2% Zn content ($x = 0.02$), the anomaly becomes less significant and moves to a lower temperature ($T = 140$ K) and the semiconductor behavior at low temperatures becomes more obvious. As $x$ increases to 0.05, this anomaly almost disappears, and the value of the resistivity increases rapidly, and resistivity becomes more semiconductor-like. The inset shows the derivative of resistivity, and as the Zn content increases, the SDW order is gradually suppressed and disappears at $x = 0.05$. For $x > 0.1$, the resistivity value increases more significantly, and at low temperature, the resistivity versus temperature is logarithmic, which should be due to the disorder effect of Zn doping which causes the localization of charge carriers. Since the zinc $3d$ electrons should not contribute to the charge carriers, the impurity scattering can lead to an increase in resistivity.

In order to study the evolution of the SDW order as a function of Zn content, we also measured the temperature dependence of magnetic susceptibility under a magnetic field of 1000 Oe, as shown in Fig. 11.1(b). A clear drop in susceptibility related to SDW order can be observed around 150 K for the parent compound, which is consistent with previous reports [30]. With increasing Zn content, the susceptibility increases gradually, and the linear temperature dependence of susceptibility for the temperature above the SDW transition, which can result from the suppression of antiferromagnetic fluctuations of $Fe^{2+}$ ions, becomes less and less significant. The Curie-like upturn observed at low temperatures can be mainly due to an extrinsic origin (such as defects and trace impurities). The inset in Fig. 11.1(b) shows enlarged plots for parent compound and 2% Zn-doped magnetic susceptibility. We can clearly observe the SDW anomaly, consistent with the resistivity result. However, no

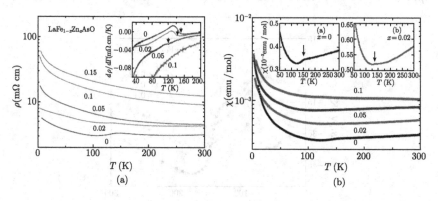

Fig. 11.1  (a) The temperature dependence of resistivity for different Zn-doped samples. The inset shows the derivative of resistivity versus temperature. (b) The temperature dependence of magnetic susceptibility. The insets show the enlarged plots for parent compound and $x = 0.02$.

anomalous change in susceptibility can be observed for $x > 0.05$. The magnetic data show that the SDW order is very sensitive to Zn doping, and only a small amount of Zn can completely destroy the SDW order.

In order to confirm further the suppression of SDW order by Zn doping, the temperature dependence of thermopower $(S)$ was also measured. The thermopower is very sensitive to the electronic structure at Fermi surface, and therefore it is an effective means to detect the SDW order. Figure 11.2 shows the temperature dependence of thermopower for a series of Zn doped samples. The thermopower of parent compounds has a sharp increase below the temperature of SDW order and structure transition, indicating that the system undergoes an abrupt change in its electronic structure, consistent with the previous report [30]. The anomaly in thermopower occurs around the structural transition, and could result from opening of a pseudogap in Fermi surface. Thus, this anomaly can be considered as a signature experimental evidence of the SDW order. It can be seen from Fig. 11.2 that the anomaly in $S$ becomes less significant and moves to a lower temperature for $x = 0.02$, still is observable for $x = 0.05$, but almost disappears for $x > 0.10$. The thermopower data confirm further that the SDW order is severely suppressed by Zn doping, and the samples become non-superconducting semiconductor-like. These experimental results show that a small amount of non-magnetic Zn impurities can severely suppress the SDW order, but do not introduce the charge carrier, and cannot

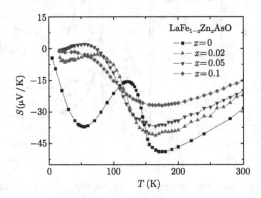

Fig. 11.2    The temperature dependence of thermopower $(S)$ for LaFe$_{1-x}$Zn$_x$AsO samples.

lead to superconductivity. This is one example of absence of superconductivity after the suppression of SDW order. Meanwhile, Zhang and Singh [27] performed the first-principles calculations of $LaFe_{1-x}Zn_xAsO$ band structure, and found that Zn is in the fully occupied state with $3d^{10}$ electronic configuration, and the Zn $3d$ states are at about $-7$ eV binding energy. There is no contribution of Zn $3d$ states to Fermi surface. Namely, the $3d$ electrons of Zn in this structure are localized, which is essentially different from the doping effect of Co, Ni and other elements. The partial substitution of Fe by Co or Ni can introduce charge carriers, and thus superconductivity emerges when the SDW order is suppressed [31−33].

### 11.3.2   Zn-Impurity Effect in the $LaFeAsO_{1-y}F_y$ Superconductors

We then studied the Zn-impurity effect on $T_c$ of the F-doped $LaFeAsO_{1-y}F_y$ superconductors with different F doping concentration. We selected $y = 0.05$, 0.1, 0.15 three concentrations as the representative samples in the underdoped, optimally doped and over-doped regimes to study the Zn-impurity effect. The XRD patterns of $LaFe_{1-x}Zn_xAsO_{1-y}F_y$ system indicate that all the samples are single phase at small $x$ ($x \leqslant 0.1$). When $x \geqslant 0.1$, the LaZnAsO foreign phase appears in the XRD diffraction pattern, resulting in phase separation. Furthermore, the lattice parameters $a$ and $c$ were obtained by fitting the diffraction peaks. The changes in both $a$- and $c$-axis become saturated for $x \geqslant 0.05$, indicating that the samples with high Zn content may lead to phase separation, and the solution of Zn becomes saturated. The Rietveld method was employed to get the Fe-As-Fe bond angle. It was found that the Fe-As-Fe bond angle in the under-doped regime changes slowly, but it decreases rapidly in the over-doped regime, which indicates that the effect of Zn impurity on superconductivity ($T_c$) could be different in different F-doping regimes.

We measured the temperature dependence of resistivity and magnetic susceptibility of the samples. Figure 11.3 shows the temperature dependence of resistivity of under-doped, optimally doped and over-doped samples from 2 K to 300 K. In the under-doped regime ($y = 0.05$), the resistivity shows an SC transition at 16.8 K ($T_c$ defined as the midpoint). To our surprise, $T_c$ even increases with increasing Zn content ($x$), $T_c$ to 22.7 K for $x = 0.06$, the metallicity of resistivity is also enhanced, and the anomaly of low-temperature resistivity (possibly from the residual SDW order) has also been suppressed

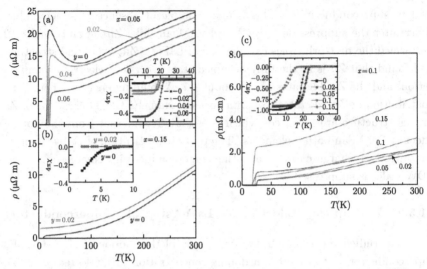

Fig. 11.3   The temperature dependence of resistivity and magnetic susceptibility of LaFe$_{1-x}$Zn$_x$AsO$_{1-y}$F$_y$ samples. The insets show the dc magnetic susceptibility versus temperature, indicating the diamagnetic shielding effect. (a) $y = 0.05$, (b) $y = 0.15$, (c) $y = 0.1$.

gradually.  The inset shows that the results of susceptibility measurement, which are consistent with resistivity measurement.  The volume fraction of superconducting magnetic shielding is nearly 95% for the samples, estimated from the magnetic susceptibility, indicating the high quality of the samples. In the optimally doped regime ($y = 0.1$), $T_c$ slightly increases from 23 K to 25.5 K and 25.4 K for $x = 0.02$ and 0.05, respectively, and the volume fraction of superconducting magnetic shielding is nearly 100% for three samples (taken into account demagnetizing factor).  For $x \geqslant 0.1$, there is a large increase in the normal state resistivity and the superconducting transition becomes broad due to the phase separation.  Meanwhile, the volume fraction of superconducting magnetic shielding also decreases quickly for $x \geqslant 0.1$, indicating the existence of non-superconducting impurity phases.  In the case of over-doped regime ($y = 0.15$), SC transition temperature ($T_c$) is 9.6 K without Zn doping.  As Zn content $x > 0.02$, superconductivity is severely suppressed, and $T_c$ decreases down to below 2 K.  The resistivity value increases slightly, and the magnetic susceptibility diamagnetic signal can no longer be observed.  In this

LaFeAs(O,F) iron-based superconducting system, for different electron-doping concentration, Zn impurity has so different effect on SC transition temperature ($T_c$), which implies that the pairing symmetry of this system is also related with the doping concentration.

In order to confirm the $3d$ electrons of Zn impurities are localized without introducing electron-type carriers, the Hall effect measurement was performed. Figure 11.4 shows the temperature dependence of the Hall coefficient ($R_H$) for under-doped and over-doped samples. For all the samples, the negative Hall coefficient ($R_H$) signal indicates that electron-type charge carriers are dominant in the whole temperature regime. For the under-doped samples, $R_H$ of the Zn-free sample exhibits a strong $T$ dependence and drops sharply below 100 K, which may be attributed to the residual SDW order or fluctuations, and similar $R_H$ behavior is particularly significant in the parent compound LaFeAsO [34]. $R_H$ finally goes to zero as the samples enter the SC state.

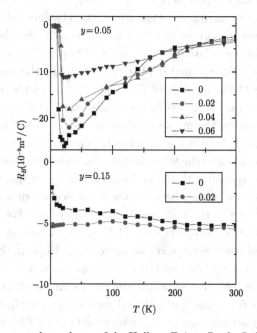

Fig. 11.4  Temperature dependence of the Hall coefficient $R_H$ for LaFe$_{1-x}$Zn$_x$AsO$_{1-y}$F$_y$, (a) $y = 0.05$, (b) $y = 0.15$.

With increasing Zn content, the sharp drop of $R_H$ is gradually suppressed, indicating that the residual SDW order or fluctuations are further suppressed by Zn impurities. Such an effect of Zn impurity on the SDW order has also been observed in the parent compound LaFeAsO. Meanwhile, the room temperature $R_H$ remains almost unchanged with Zn doping, indicating that Zn doping does not change the charge carrier density. This is consistent with the band calculation result [27]. For the over-doped samples, $R_H$ of all Zn-doped samples exhibits very weak $T$ dependence and again the change in the room temperature $R_H$ is very small, indicating that the charge carrier density essentially does not change with Zn doping. As temperature decreases, $R_H$ drops quickly to zero due to SC transition for the Zn-free sample, but for the Zn-doped ($x = 0.02$) samples, $T_c$ is suppressed to below 2 K and $R_H$ remains constant, exhibiting a typical metallic behavior.

To further understand the effect of Zn doping, we also measured temperature dependence of the thermopower for these samples, as shown in Fig. 11.5. For under-doped samples ($y = 0.05$), as the Zn content increases to 0.04, $T_c$ gradually increases. Meanwhile, the absolute value of its normal state thermopower also gradually increases. Actually the thermopower values of all samples are relatively large, up to 120 $\mu$V/K, consistent with the other reports [34]. For optimal doping ($y = 0.1$), the thermopower absolute values increase with the increase of $T_c$. When $T_c$ reaches the maximum value, the room temperature thermopower absolute value is also largest. In fact, the room temperature thermopower absolute values in the SC phase of the FeAs-based superconductors are significantly greater than that of the parent compounds, in contrast to the SC cuprates. The enhanced thermopower in the SC phase may be related with the multi-band effect or magnetic fluctuations. There may be a relationship between the enhanced thermopower and high-$T_c$ superconductivity in the iron-based superconductors. For the over-doped samples ($y = 0.15$), as shown in Fig. 11.5(c), as Zn content is beyond 0.02, $T_c$ disappears and the thermopower absolute value drops quickly, which again indicates that $T_c$ is correlated with the thermopower value.

We summarize the above experimental results on the effect of Zn impurity on $T_c$ in Fig. 11.6. In the under-doped regime ($y = 0.05$), with increasing Zn content, $T_c$ gradually increases to a maximum at 22.7 K. In the optimally doped regime ($y = 0.1$), $T_c$ remains almost unchanged, even slightly increases.

On the contrary, in the over-doped regime ($y = 0.15$), $T_c$ drops sharply, and only 0.02 Zn impurity can completely suppress $T_c$. According to the results of Hall effect measurements, in both under-doped and over-doped regime, Zn impurities will not change the carriers concentration, and Zn impurities have different effect on $T_c$ for different F-doping concentration, which may indicate that superconducting pairing symmetry will change with F-doping concentration.

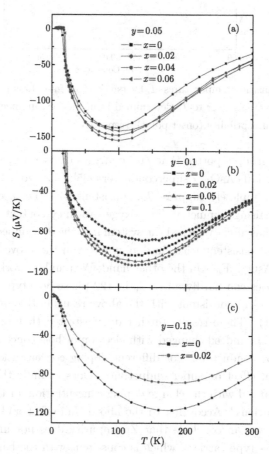

Fig. 11.5  The temperature dependence of thermopower ($S$) for LaFe$_{1-x}$Zn$_x$AsO$_{1-y}$F$_y$ samples, (a) $y = 0.05$, (b) $y = 0.1$, (c) $y = 0.15$.

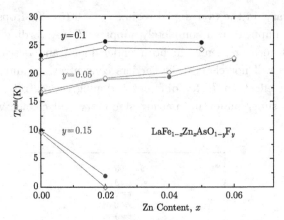

Fig. 11.6   SC transition temperatures ($T_c$) versus Zn content in LaFe$_{1-x}$Zn$_x$AsO$_{1-y}$F$_{\it y}$
Solid and open symbols refer to $T_c$ determined from the measurements of resistivity
(midpoint) and susceptibility (onset point), respectively.

A Japanese group reported that Zn impurities can severely suppress $T_c$ in
oxygen deficient LaFeAsO$_{1-\delta}$ superconductors [35]. They found that only 3%
Zn doping can completely suppress $T_c$, consistent with the theoretical model
for $s+-$ wave pairing symmetry. The oxygen vacancy content is 0.2 in their
samples, which has already turned the system into the over-doped regime, thus
their results are consistent with Zn-doping effect in the F-overdoped regime
in LaFe$_{1-x}$Zn$_x$AsO$_{1-y}$F$_y$. On the other hand, Wen and co-workers confirmed
that Zn impurities can hardly affect $T_c$ in 122 phase hole-type superconductors Ba$_{1-x}$K$_x$Fe$_2$As$_2$, consistent with the above result in the optimally doped
regime ($y = 0.1$). These results further demonstrated that the effect of Zn
impurities could be indeed changed with electron or hole concentration in the
FeAs-based superconductors. For different doping concentrations, Zn impurity has different effect on superconductivity. Does it imply that SC pairing
symmetry is changed with the charge doping concentration, or there are other
possible explanations? According to the above Hall effect and thermopower
measurements, we can conclude that Zn impurity does not introduce extra
electron- or hole- type carriers, which is consistent with the band calculation
results. In this way, the most likely explanation about the Zn-impurity effect
is that the pairing symmetry of iron-based superconductors is not universal,
but it may change with the doping concentration (i.e. electronic structure).

Kuroki *et al.* [36] have used a five-band model to study the superconducting pairing symmetry mediated by spin fluctuations, and found that the distance ($hp$) from As or P to the FeAs plane can affect the pairing symmetry, namely, from high $T_c$, nodeless SC gap to low $T_c$ nodal SC gap. From LaFePO to LaFeAsO, then to NdFeAsO, the height $hp$ gradually increases, and its superconducting transition temperature also gradually increases (NdFeAsO$_{1-y}$, $T_c = 52$ K). That is, at the higher $hp$ value, its $T_c$ is also higher, and the SC gap has no node, while at smaller $hp$ value, such as the LaFePO system, $T_c$ is lower and SC gap could have linear nodes. Moreover, electron or hole-type doping can change the value of $hp$, and thus it is proposed that the pairing symmetry can change at different doping concentration.

We believe that in the under-doped and over-doped regimes, the opposite effect of Zn impurity on $T_c$ should result from the change in SC pairing symmetry, and the different Zn-impurity effect cannot be explained by a single pairing symmetry. In the under-doped regime, the SC pairing symmetry may be $s$-wave ($s++$) pairing, but it may change to $s+-$ or $d$-wave pairings with nodal energy gap in the over-doped regime. The iron-based SC compounds have electron and hole Fermi pockets, the $s$-wave and $s+-$ wave states correspond to the same and opposite signs of the relative order parameters between the hole and electron Fermi pockets, respectively. It could be expected that from the under-doped regime to over-doped regime the change in the Josephson coupling between the Fermi pockets will lead to the switch of the pairing symmetry from $s++$ to $s+-$ wave. The impurity effect in the $s+-$ wave state was studied by Bang *et al.* [21] and by Onari *et al.* [22]. Theoretical calculations have shown that, in the strong scattering limit, the non-magnetic impurity effect on the $s+-$ wave state is severe and similar to the effect on the $d$-wave SC state. Therefore, we believe that the system is $s+-$ or $d$-wave pairing in the over-doped regime. Meanwhile, the ARPES experiments [13–14] have found that the SC gap is isotropic, ruling out the possibility of $d$-wave pairing in the over-doped regime. Thus it may be $s+-$ pairing. Obviously, the results of Zn impurity effect in the under-doped regime are not consistent with the $s+-$ wave pairing symmetry, therefore it should be $s$-wave ($s++$ wave) pairing symmetry. In our experiments, $T_c$ increases with increasing Zn content, which could be due to destruction of the residual SDW order by Zn impurity. In the under-doped regime, the SDW order competes with super-conducting order, and Zn impurity suppresses the SDW order, and it has no effect on the $s$-wave pairing symmetry, thus $T_c$ increases. Therefore, we believe that the SC pairing symmetry will change in the F-doped 1111 phase superconductors, as F-doping concentration changes.

Other experiments also support the above conclusions. For example, Taillefer's group [37] studied the low temperature thermal conductivity of 122 phase BaFe$_{2-x}$Co$_x$As$_2$. They proposed that the superconducting order parameter changes from $s$-wave to $d$-wave with increasing Co content, which provides a possible example that the pairing symmetry of iron-based superconductors evolves with electron-doping concentration. For Ni-doped Ba(Fe$_{1-x}$Ni$_x$)$_2$As$_2$ superconductors, the magnetic penetration depth experiment along $c$-axis [38] found that the low temperature magnetic penetration remains almost constant, and its value tends to zero in the under-doped regime ($x = 0.033$), indicating almost no quasi-particle excitations. It is consistent with $s$-wave behavior which has isotropic energy gap. But in the over-doped regime ($x = 0.072$), the magnetic penetration depth varies linearly with temperature, indicating that the SC energy gap has a node or nodal line, matching the $d$-wave symmetry. In addition, the NMR experiments on BaFe$_2$As$_{2-x}$P$_x$ [39] also supports the change in SC pairing symmetry with the change of doping concentration.

Incidentally, in Co-doped 1111 phase superconductors LaFe$_{1-x}$Co$_x$AsO, in both under-doped and over-doped regimes, Zn impurity has significantly suppressed superconductivity [40]. Other groups also obtained similar Zn-impurity effect in the Co-doped 122 phase superconductors BaFe$_{2-x}$Co$_x$As$_2$ [41]. These results indicate that SC pairing symmetry in the Co-doped system may be $d$-wave or $s+-$ wave, and it also suggests that even in the same 1111 phase iron-based superconductors, the SC pairing symmetry may also be different due to the different doping methods.

## 11.4    Conclusion

In summary, we have systematically studied the effect of a non-magnetic Zn impurity on superconductivity in the 1111 phase iron-based superconductors. We found that only a small amount of Zn can completely suppress SDW order in the parent compounds, and the system becomes semiconductor-like from semimetal-like. In the optimally-doped superconductors, slight Zn doping does not affect $T_c$, and even help to enhance $T_c$. Subsequently, the Zn-doping effect for the different electron-doping concentration was further investigated, and it was found that Zn doping can destroy the residual SDW order to enhance $T_c$ in the under-doped regime ($y = 0.05$), but it severely suppresses $T_c$ in the over-doped regime. Different Zn-doping effect for different electron doped regimes may come from the switch of superconducting pairing symmetry. In the under-doped regime (0.05), the SC symmetry is $s$-wave, insensitive to non-magnetic impurity. However the symmetry may be changed to $s+-$ or $d$-wave in the over-doped regime (0.15), and the non-magnetic impurity has a great impact on $T_c$. Based on our experimental results, we propose that the SC pairing symmetry could change with the electron doping concentration.

## Acknowledgments

This work is supported by the Natural Science Foundation of Zhejiang Province (grant no. Y6100216), the Natural Science Foundation of China (grant no. 10931160425 and 11174247), and the National Basic Research Program of China (grant no. 2011CBA00103).

---

[1] Y. Kamihara, T. Watanabe, M. Hirano and H. Hosono, J. Am. Chem. Soc. **130**, 3296 (2008).

[2] X. H. Chen, T. Wu, G. Wu, R. H. Liu, H. Chen and D. F. Fang, Nature **453**, 761 (2008).

[3] G. F. Chen, Z. Li, D. Wu, G. Li, W. Z. Hu, J. Dong, P. Zheng, J. L. Luo and N. L. Wang, Phys. Rev. Lett. **100**, 247002 (2008).

[4] Z. A. Ren, W. Lu, J. Yang, W. Yi, X. L. Shen, Z. C. Li, G. C. Che, X. L. Dong, L. L. Sun, F. Zhou and Z. X. Zhao, Chin. Phys. Lett. **25**, 2215 (2008).

[5] H. H. Wen, G. Mu, L. Fang, H. Yang and X. Zhu, Europhys. Lett. **83**, 17009 (2008).

[6] C. Wang, L. Li, S. Chi, Z. Zhu, Z. Ren, Y. Li, Y. Wang, X. Lin, Y. Luo, S. Jiang, X. Xu, G. Cao and Z. Xu, EPL **83**, 67006 (2008).

[7] S. Takeshita, R. Kadono, M. Hiraishi, M. Miyazaki, A. Koda, S. Matsuishi and H. Hosono, Phys. Rev. Lett. **103**, 027002 (2009).

[8] M. Rotter, M. Tegel, D. Johrendt, I. Schellenberg, W. Hermes and R. Pttgen, Phys. Rev. B. **78**, 020503 (2008).

[9] X. C. Wang, Q. Q. Liu, Y. X. Lv, W. B. Gao, L. X. Yang, R. C. Yu, F. Y. Li and C. Q. Jin, Solid State Communications **148**, 538 (2008).

[10]  F. C. Hsu, J. Y Luo, K. W. Yeh, T. K. Chen, T. W. Huang, P. M. Wu, Y. C. Lee, Y. L. Huang, Y. Y. Chu, D. C. Yan and M. K. Wu, Proc Natl Acad Sci 105, 14262, (2008).

[11]  J. G. Guo, S. F. Jin, G. Wang, S. C. Wang, K. X. Zhu, T. T. Zhou, M. He and X. L. Chen, Phys. Rev. B 82, 180520 (2010).

[12]  M. H. Fang, H. D. Wang, C. H. Dong, Z. J. Li, C. M. Feng, J. Chen and H. Q. Yuan, EPL 94, 27009 (2011).

[13]  H. Ding, P. Richard, K. Nakayama, K. Sugawara, T. Arakane, Y. Sekiba, A. Takayama, S. Souma, T. Sato, T. Takahashi, Z. Wang, X. Dai, Z. Fang, G. F. Chen, J. L. Luo and N. L. Wang, EPL 83, 47001 (2008).

[14]  T. Kondo, A. F. S. Syro, O. Copie, C. Liu, M. E. Tillman, E. D. Mun, J. Schmalian, S. L. Bud ko, M. A. Tanatar, P. C. Canfield and A. Kaminski, Phys. Rev. Lett. 101, 147003 (2008).

[15]  K. Hashimoto, T. Shibauchi, T. Kato, K. Ikada, R. Okazaki, H. Shishido, M. Ishikado, H. Kito, A. Iyo, H. Eisaki, S. Shamoto and Y. Matsuda, Phys. Rev. Lett. 102, 017002 (2008).

[16]  K. Hashimoto, T. Shibauchi, S. Kasahara, K. Ikada, S. Tonegawa, T. Kato, R. Okazaki, C. J. van der Beek, M. Konczykowski, H. Takeya, K. Hirata, T. Terashima and Y. Matsuda, Phys. Rev. Lett. 102, 207001 (2009).

[17]  H. J. Grafe, D. Paar, G. Lang, N. J. Curro, G. Behr, J. Werner, J. H. Borrero, C. Hess, N. Leps, R. Klingeler and B. Buchner, Phys. Rev. Lett. 101, 047003 (2008).

[18]  H. Fukazawa, T. Yamazaki, K. Kndo, Y. Konori, N. Takeshita, P. M. Shirage, K. Kihou, K. Miyazawa, H. Kito, H. Eisaki and A. Iyo, J. Phys. Soc. Jpn. 78, 033704 (2009).

[19]  I. I. Mazin, D. J. Singh, M. D. Johannes and M. H. Du, Phys. Rev. Lett. 101, 057003 (2008).

[20]  V. Cvetkovic and Z. Tesanovic, EPL 85, 37002 (2009).

[21]  Y. Bang, H. Choi and H. Won, Phys. Rev. B 79, 054529 (2009).

[22]  S. Onari and H. Kontani, Phys. Rev. Lett. 103, 177001 (2009).

[23]  P. W. Anderson, J. Phys. Chem. Solids 11, 26 (1959).

[24]  G. Xiao, M. Z. Cieplak, A. Gavrin, F. H. Streitz, A. Bakhshai and C. L. Chien, Phys. Rev. Lett. 60, 1446 (1988).

[25]  G. Xiao, A. Bakhshai, M. Z. Cieplak, Z. Tesanovic and C. L. Chien, Phys. Rev. B 39, 315 (1989).

[26]  K. Ishida, Y. Kitaoka, N. Ogata, T. Kamino, K. Asayama, J. R. Cooper and N. Athanassopoulou, J. Phys. Soc. Jpn 62, 2803 (1993).

[27]  L. J. Zhang and D. J. Singh, Phys. Rev. B 80, 214530 (2009).

[28]  Y. K. Li, X. Lin, Q. Tao, C. Wang, T. Zhou, L. J. Li, Q. B. Wang, M. He, G. H. Cao and Z. A. Xu, New J. Phys. 11, 053008 (2009).

[29]  Y. K. Li, J. Tong, Q. Tao, C. M. Feng, G. H. Cao, W. Q. Chen, F. C. Zhang and Z. A. Xu, New J. Phys. **12**, 083008 (2010).

[30]  M. A. McGuire, A. D. Christianson, A. S. Sefat, B. C. Sales, M. D. Lumsden, R. Jin, E. A. Payzant, D. Mandrus, Y. Luan, V. Keppens, V. Varadarajan, J. W. Brill, R. P. Hermann, M. T. Sougrati, F. Grandjean and G. J. Long, Phys. Rev. B. **78**, 094517 (2008).

[31]  A. S. Sefat, A. Huq, M. A. McGuire, R. Y. Jin, B. C. Sales, D. Mandrus, L. M. D. Cranswick, P. W. Stephens and K. H. Stone, Phys. Rev. B **78**, 104505 (2008).

[32]  C. Wang, Y. K. Li, Z. W. Zhu, S. Jiang, X. Lin, Y. K. Luo, S. Chi, L. J. Li, Z. Ren, M. He, H. Chen, Y. T. Wang, Q. Tao, G. H. Cao and Z. A. Xu, Phys Rev B **79**, 054521 (2009).

[33]  L. J. Li, Y. K. Luo, Q. B. Wang, H. Chen, Z. Ren, Q. Tao, Y. K. Li, X. Lin, M. He, Z. W. Zhu, G. H. Cao and Z. A. Xu, New J. Phys. **11**, 025008 ( 2009).

[34]  M. A. McGuire, A. D. Christianson, A. S. Sefat, B. C. Sales, M. D. Lumsden, R. Jin, E. A. Payzant, D. Mandrus, Y. Luan, V. Keppens, V. Varadarajan, J. W. Brill, R. P. Hermann, M. T. Sougrati, F. Grandjean and G. J. Long, Phys. Rev. B. **78**, 094517 (2008).

[35]  Y. F. Guo, Y. G. Shi, S. Yu, A. A. Belik, Y. Matsushita, M. Tanaka, Y. Katsuya, K. Kobayashi, I. Nowik, I. Felner, V. P. S. Awana, K. Yamaura and E. Takayama-Muromachi, Phys. Rev. B **82**, 054506 (2010).

[36]  K. Kuroki, H. Usui, S. Onari, R. Arita and H. Aoki, Phys. Rev. B **79** 224511, (2009).

[37]  M. A. Tanatar, J. P. Reid, H. Shakeripour, X. G. Luo, N. Doiron-Leyraud, N. Ni, S. L. Bud'ko, P. C. Canfield, R. Prozorov and L. Taillefer, Phys. Rev. Lett. **104**, 067002, (2010).

[38]  C. Martin, H. Kim, R. T. Gordon, N. Ni, V. G. Kogan, S. L. Bud'ko, P. C. Canfield, M. A. Tanatar and R. Prozorov, Phys. Rev. B **81**, 060505, (2010).

[39]  Y. Nakai, T. Iye, S. Kitagawa, K. Ishida, S. Kasahara, T. Shibauchi, Y. Matsuda and T. Terashima, Phys. Rev. B **81**, 020503(R) (2010).

[40]  Y. K. Li, J. Tong, Q. Tao, G. H. Cao, Z. A. Xu, Journal of Physics and Chemistry of Solids **72**, 410 (2011).

[41]  J. Li, Y. F. Guo, S. B. Zhang, S. Yu, Y. Tsujimoto, H. Kontani, K. Yamaura and E. Takayama-Muromachi, Phys. Rev. B **84**, 020513 (2011).

# 12
# Very-Low-Temperature Heat Transport of High-Temperature Superconductors

X. F. Sun

*University of Science and Technology of China*

It has been more than 20 years since the high-temperature superconductors (HTSCs), including cuprate and iron-based superconductors, were discovered. The various physical properties of the normal state and the superconducting state have been studied in great details. Although the intrinsic mechanism of superconductivity has not been revealed, a variety of experiments indicate that these materials present distinct characteristics compared with conventional $s$-wave symmetrical superconductors, and therefore, these materials are considered as unconventional superconductors. The nature and transport behaviors of the low-energy quasiparticles (QPs) in the superconducting state are undoubtedly important to explore the unconventional superconducting electronic states and the mechanism of superconductivity. Among kinds of experimental methods, the very-low-temperature thermal conductivity ($\kappa$) plays a very important role. This article mainly introduces some experimental progress on the very-low-temperature heat transport of these two types of unconventional superconductors, including the universal thermal conductivity behavior, the superconducting gap symmetry reflected by the residual thermal conductivity, the very-low-$T$ thermal conductivity and the metal-insulator transition of the normal state in copper oxide superconductors.

## 12.1  Introduction of Heat Transport

The thermal conductivity is related to the ability of energy transport of elementary excitations in solid, including phonons, electrons or magnons, etc.

The classical kinetic theory gives a simple formula [1]

$$\kappa = \frac{1}{3}Cvl, \tag{12.1}$$

where $C$ is the heat capacity, $v$ is the mean velocity, and $l$ is the mean free path of heat carrier. Thermal conductivity can provide the physical information of a variety of elementary excitations. When there are more than one kind of heat carriers in the system, the behavior of thermal conductivity usually becomes complicated and difficult to analyze.

Phonon thermal conductivity exists in all solid materials and its temperature dependence is very complicated. Only at very low temperatures (usually below 10 K in the insulating materials, and at even lower temperatures in the cuprates) the phonon thermal conductivity $\kappa_{ph}$ shows a simple $T^3$ relation. This is because at very low temperatures, the microscopic scattering mechanisms like phonon-phonon scattering and defect scattering are smeared out, the mean free path of phonon is gradually increased and finally determined by the geometric size of the sample and is independent of the temperature. This is the so-called boundary scattering limit. The $T^3$ relation of $\kappa(T)$ results from the temperature dependence of the phonon specific heat at low temperatures. Such a kind of temperature dependence is helpful for separating the electronic thermal conductivity from the phononic one in metals or unconventional superconductors at very low temperatures so as to obtain QP (electron) thermal conductivity $\kappa_{el}$ we are more concerned.

For the conventional $s$-wave superconductors, the study on low-$T$ heat transport is not very informative. The reason is very simple, that is, the $s$-wave superconducting gap in the momentum space is isotropic, so the number of QP excitations decreases exponentially with decreasing temperature in the superconducting state [2]. Therefore, the contribution of QPs to the thermal conductivity at very low temperatures is negligibly small and can provide us little information.

For an unconventional superconductor with $p$-wave, $d$-wave or other symmetries, the superconducting gaps along some particular directions in the momentum space are zero, the so-called gap nodes. The low-energy QPs can be excited and form a finite density near the nodes as long as the impurity scattering works. These QPs can exist at very low temperatures and contribute to the heat transport, so the low-$T$ thermal conductivity of HTSC could include a significant contribution from QPs [3]. But there is a problem: how to separate the contribution of electrons from the phonon conductivity? It is usually easy to achieve for the data at very low temperatures. As mentioned above, $\kappa_{\mathrm{ph}}$ is proportional to $T^3$ and $\kappa_{\mathrm{el}}$ is linear with $T$ at very low temperatures, one can get

$$\frac{\kappa}{T} = a + bT^2. \tag{12.2}$$

As shown in Fig. 12.1, the plot of $\kappa/T$ vs $T^2$ is expected to be well fitted linearly at very low temperatures, with the slope $b$ characterizing the phonon contribution and the intercept $a$ the contribution of electrons, which is named as the residual thermal conductivity (often written as $\kappa_0/T$). The first important issue of the very-low-$T$ thermal conductivity is to study whether there is non-zero $\kappa_0/T$, because it is one of the most effective ways to judge whether the order parameter is the traditional $s$ wave or the unconventional $d$ wave (or $p$ wave) with nodes. Second, the dependencies of $\kappa_0/T$ on the concentration of heat carriers and the magnetic field also provide physics about the QP transport, the ground-state properties, and the quantum phase transitions, etc. Since equation (12.2) is usually established only at very low temperatures, the very-low-$T$ measurement becomes an indispensable tool to study the QP heat transport in HTSC.

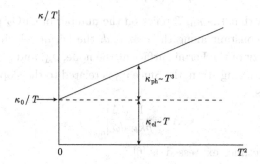

Fig. 12.1   At very-low-temperature region, in which the boundary scattering limit establishes, $\kappa/T$ is proportional to $T^2$. The slope is dependent on the phonon thermal conductivity and the residual thermal conductivity $\kappa_0/T$ is the contribution of electrons.

## 12.2    Thermal Conductivity of Cuprate HTSC and $d$-Wave Sym metry

Universal thermal conductivity is a prediction by the Fermi-liquid theory on the classical $d$-wave superconductors [4,5]. For the quasi-2D superconductors with $d_{x^2-y^2}$ symmetry, the anisotropic energy gap can be expressed as

$$\Delta(k) = \Delta_0 \cos 2\phi, \tag{12.3}$$

where $\Delta_0$ is the maximum of the energy gap, $\phi$ is the in-plane azimuth angle in the momentum space. Due to the impurity scattering, there could always be low-energy QP excitations near the nodes. Although the impurities can create the elementary excitations, meanwhile they also scatter QPs. In some cases, the effects of creation and scattering balance, causing the thermal conductivity of QPs to be independent on the strength of impurity scattering. This phenomenon is called the universal thermal conductivity. By using the self-consistent $T$-matrix approximation (SCTMA) theory for the classical $d$-wave superconductors, one can get that, when the impurity scattering strength meets $k_B T \ll \gamma \ll \Delta_0$ ($\gamma$ is the width of impurity energy band), the residual thermal conductivity is independent of the strength of QP scattering, and satisfies the following relation [5]

$$\frac{\kappa_0}{T} = \frac{k_B^2}{3\hbar} \frac{n}{d} \left( \frac{v_F}{v_\Delta} + \frac{v_\Delta}{v_F} \right) \approx \frac{k_B^2}{3\hbar} \frac{n}{d} \frac{v_F}{v_\Delta}. \tag{12.4}$$

It is clearly seen that the $\kappa_0/T$ relies on the number of $CuO_2$ plane $n$ in unit cell, the lattice constant along the $c$ axis, $d$, the Fermi velocity, $v_F$, and the tangential velocity of the Fermi surface at the node, $v_\Delta$, and is independent of the impurity scattering strength. Since $v_\Delta$ is related to the slope of the energy gap at the nodes

$$v_\Delta = \frac{1}{\hbar k_F} \frac{d\Delta}{d\phi}\bigg|_{node}, \tag{12.5}$$

the $\kappa_0/T$ is sometimes expressed as [6]

$$\frac{\kappa_0}{T} \approx \frac{k_B^2}{6\hbar} \frac{n}{d} \frac{k_F v_F}{\Delta_0}. \tag{12.6}$$

From the experimental results of the angle-resolved photoelectron spectroscopy (ARPES), we know that $k_F$ and $\nu_F$ are the parameters that have little to do with the materials and the carrier concentrations, so the $\kappa_0/T$ is directly related to the superconducting gap $\Delta_0$. The universal thermal conductivity is a criterion of whether the HTSC can be described by the classical $d$-wave theory. If the scenario is correct, it would be very helpful to provide a true bulk measurement of superconducting gap, which is essentially more effective than the ARPES and tunneling spectroscopy experiments.

Shortly after the theoretical prediction of the universal thermal conductivity, Taillefer $et$ $al.$ gave the first experimental evidence [7]. As shown in Fig. 12.2, they observed the non-zero $\kappa_0/T = 0.019$ W/(K²·m) in the optimally doped $YBa_2Cu_3O_y$ (YBCO, $y = 6.9$). By doping different amounts of Zn to change the impurity scattering strength and measuring the thermal conductivity, the authors pointed out that the $\kappa_0/T$ of the samples with Zn content ranging from 0% to 3% is nearly constant [$\kappa_0/T$=(0.017~0.025) W/(K²·m)]. They concluded that the residual thermal conductivity is independent of impurity scattering strength, which directly confirmed the picture of the universal thermal conductivity.

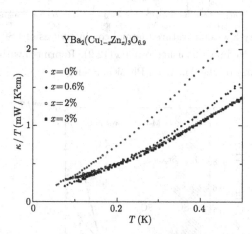

Fig. 12.2   The $a$-axis thermal conductivities of four Zn-doped YBCO crystals, plotted as $\kappa/T$ vs $T$. Reprinted with permission from [7]. Copyright (1997) by the American Physical Society.

In fact, subsequent studies found that for a variety of HTSC, there was a "universal" dependence of $\kappa_0/T$ on the carrier concentration. Takeya $et$ $al.$ [8] systematically studied the relationship between the $\kappa_0/T$ and the carrier concentration for the first time. They measured the thermal conductivity of a series of $La_{2-x}Sr_xCuO_4$ (LSCO) single crystals with $x = 0 \sim 0.22$ down to tens of mK, as shown in Fig. 12.3. Using formula (12.2) to separate the $\kappa_0/T$ term from the experimental data, they found that $\kappa_0/T$ in the non-superconducting samples with $x \leqslant 0.05$ are nearly zero. The $\kappa_0/T$ becomes finite near $x = 0.06$, accompanied with the presence of superconductivity. Then, except for the singularity at $x = 1/8$, $\kappa_0/T$ in all the other samples increased with $x$, as shown in Fig. 12.4.

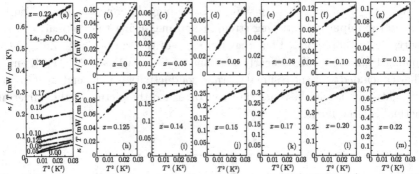

Fig. 12.3  The very-low-temperature thermal conductivities of LSCO ($x = 0 - 0.22$) single crystals and the linear fit using formula (12.2). Reprinted with permission from [8]. Copyright (2002) by the American Physical Society.

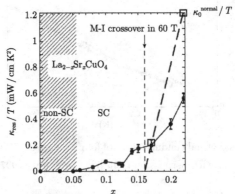

Fig. 12.4  The dependence of residual thermal conductivity on the carrier concentration in LSCO single crystals. Reprinted with permission from [8]. Copyright (2002) by the American Physical Society.

The similar behavior has also been observed in YBCO, Bi2201, Tl2201, Bi2212 and other systems. Figures 12.5 and 12.6 are the YBCO results reported by Sutherland *et al.* [9] and Sun *et al.* [10], respectively. For the sake of avoiding the contribution of electron thermal conductivity along the Cu-O chains in the $b$ axis, all these measurements were carried out along the $a$ axis of the untwined single crystals. The obtained electron thermal conductivity was completely the QP conduction behavior in the $CuO_2$ superconducting layers. These data indicated that when the carrier concentration increased with increasing the oxygen content, $\kappa_0/T$ was also gradually increased. Figure 12.7 is for the La-doped $Bi_2Sr_{2-x}La_xCuO_{6+\delta}$ (BSLCO) [11], in which the highest superconducting transition temperature could achieve above 36 K. Obviously, the relation between the $\kappa_0/T$ and doping concentration in BSLCO system is qualitatively the same as those in LSCO and YBCO.

It should be noted that there are also some experimental results that are not consistent with the universal thermal conductivity behavior [8, 12, 13]. As a matter of fact, the experimental phenomena in HTSC which are inconsistent with the universal thermal conductivity are not difficult to understand. The universal thermal conductivity is established under a simple theoretical framework, the object is a classical $d$-wave superconductor. From the view of experiments, recent scanning tunneling microscope (STM) experiments showed that the electronic states of HTSC is strongly inhomogeneous [14−20], which was

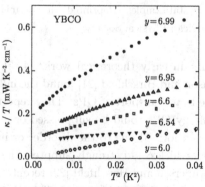

Fig. 12.5   The $a$-axis thermal conductivities of YBCO single crystals for different oxygen contents ($y$) at low temperatures. Reprinted with permission from [9]. Copyright (2003) by the American Physical Society.

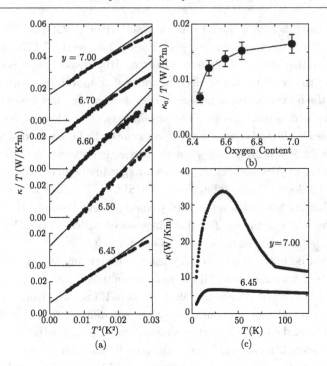

Fig. 12.6 (a) The $a$-axis thermal conductivities of YBCO single crystals for different oxygen contents ($y$) at low temperatures. (b) The $y$ dependence of residual thermal conductivity. (c) The high-temperature thermal conductivity data indicating the high quality of these single-crystal samples. Reprinted with permission from [10]. Copyright (2004) by the American Physical Society.

not taken into account in early theoretical works [4, 5]. From the view of theory, in 2002 Atkinson and Hirschfeld [21] found the influence of disorder or inhomogeneity on the heat transport of QPs. They reported that the density of state of the low-energy QPs could be suppressed or a small pseudogap at the Fermi level was even formed due to the disorder or inhomogeneity in the superconducting state, which could strongly suppress the conduction of QPs. On the other hand, Andersen and Hirschfeld [22] recently considered that the introduction of non-magnetic impurities would affect the magnetic properties of the underdoped samples and meanwhile strongly suppress the transport of QPs.

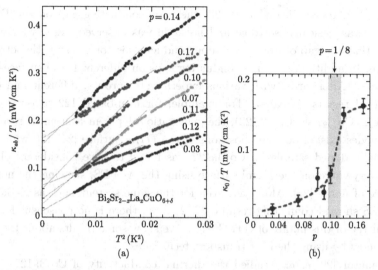

Fig. 12.7   (a) The very-low-$T$ thermal conductivities of $Bi_2Sr_{2-x}La_xCuO_{6+\delta}$ single crystals for different carrier concentrations ($p$); (b) The $p$ dependence of residual thermal conductivity. Reprinted with permission from [11]. Copyright (2004) by the American Physical Society.

In short, the low-energy QP heat transport was definitely observed in cuprate HTSC and most of them showed that the residual thermal conductivity $\kappa_0/T$ was finite, and $\kappa_0/T$ becomes larger with increasing the carrier concentration. The existence of these low-energy QPs is undoubtedly the strongest evidence to support the unconventional superconducting symmetry with gap nodes, and the presence of nodal QPs called for the new theory of the electronic states of HTSC. Particularly in the underdoped region, whether there are nodal excitations in a strongly correlated system, which has the spin/charge stripe phase and singular normal-state properties like the pseudo-gap, is an essential issue in theory.

## 12.3   The Relationship between the Residual Thermal Conductivity and Superconducting Gap in Iron-Based Superconductors

The discovery of iron-based superconductors has aroused a new upsurge of superconductivity research. The earliest study of energy gap for iron-based superconductors indicated that [23] the superconducting gap on the Fermi surface was basically isotropic and no node was observed. It is widely believed that the gap of iron-based superconductors is consistent with isotropic $s$ wave. In addition, the superconducting gaps on different Fermi surfaces are different. Afterwards, with various experimental probes, different and controversial results appeared. For example, for iron-based 122 phase system $Ba(Fe_{1-x}Co_x)_2As_2$ (Co-Ba122), the penetration depth measurement revealed that nodes existed on some positions of the Fermi surface [24, 25], while the optimally doped samples of Co-Ba122 was found to have considerably large gap everywhere on the Fermi surface using the ARPES, showing the nonexistence of nodes [26]. Moreover, even for the same system, such as K-Ba122, ARPES showed conflicting results [27, 28]. So, there is no consensus for the symmetry of this family of HTSC. Here, we show some results about the gap symmetry by using the heat transport technique.

Tanatar [29] $et$ $al.$ studied the thermal conductivity of Co-Ba122 single crystals with various doping from the underdoped to the overdoped region (from $x = 0.048$ to 0.114) and found that within the experimental error bars, all Co-Ba122 samples exhibited negligible residual linear terms, irrespective of the doping level, as shown in Fig. 12.8. This indicated that the superconducting gaps of Co-Ba122 are nodeless everywhere on the Fermi surface. By comparison, in a $d$-wave superconductor like $Tl_2Ba_2CuO_{6+\delta}$ (Tl-2201) [30], $\kappa_0/T = 0.35\kappa_N/T$, while in all four Co-Ba122 samples, $\kappa_0/T < 0.01\kappa_N/T$. This rules out the $d$-wave symmetry. At low doping, the variation of $\kappa_0/T$ exhibited an isotropic $s$-wave gap, with coexisting AFM and orthorhombic order. At high doping, the symmetry comparable to a $d$-wave gap was observed. From Fig. 12.9, the fast initial rise of $\kappa_0/T$ with $H$ showed that the gap must be very small on some portion of the Fermi surface.

The authors pointed out that in the $s$-wave symmetry the above results can be understood using the pronounced relationship between the antiferromagnetic wavevector and the gap [31–33]. Whereby the gap is $s$ wave but highly anisotropic, with zero value in some directions. In addition, the gap may also change sign, showing $s\pm$-wave pairing, namely, the superconducting order parameter has opposite signs on different Fermi surfaces, which differs from the early results [23].

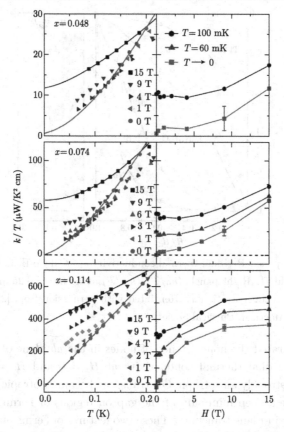

Fig. 12.8   Left panels: Temperature dependencies of in-plane thermal conductivity for Co-Ba122 crystals with different Co concentrations. The lines are a power-law fit of the form $\kappa/T = a + bT^{\alpha-1}$ ($\alpha \approx 1-1.5$). Right panels: Magnetic field dependencies of $\kappa/T$ for different Co concentrations under different temperatures. Reprinted with permission from [29]. Copyright (2010) by the American Physical Society.

Reid *et al.* [34] studied the low temperature thermal conductivity along the $c$ axis of Co-Ba122 with Co doping concentration from 0.038 to 0.127. The results showed that there existed nodes in the superconducting gap from underdoped to overdoped region, as shown in Fig. 12.10. Besides, the nodes were located on regions of Fermi surface that dominated the $c$-axis transport and contributed very little to the in-plane transport. The previous results

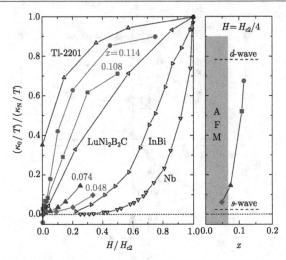

Fig. 12.9   Left panel: Residual thermal conductivity $\kappa_0/T$ of Co-Ba122 as a function of magnetic field $H$. Right panel: $(\kappa_0/T)/(\kappa_N/T)$ at $H/H_{c2} = 0.25$ in Co-Ba122 as a function of Co concentration $x$. Reprinted with permission from [29]. Copyright (2010) by the American Physical Society.

have demonstrated the nonexistence of nodes in the $ab$ plane of this system. Comparing residual thermal conductivity at $H = 0$ and $H = H_{c2}/4$, the fact that the strongly anisotropic QP transport became isotropic showed that there must be a deep minimum of the gap on regions of Fermi surface that dominates the in-plane transport. These two features of Fermi surface, that is, the nodes on 3D regions and minima on 2D regions of the Fermi surface, also pointed to the relationship between the gap $\Delta$ and $k$, which would be present on all Fermi surfaces, but is most pronounced along $k_z$. This suggested a close relation between the Fermi surface and the gap.

Different from other doped iron arsenic superconductors with node-free gap [26, 28, 34−39], the 122 phase $KFe_2As_2$, with K completely replacing Ba, does not has the $s\pm$-wave superconducting gap [40, 41]. This material has clear nodes similar to the $d$-wave gap. It is worth noting that the Fermi-surface nesting and inter-band interaction still exist in nodal superconductors such as LaFePO and $BaFe_2(As_{1-x}P_x)_2$. The main reason was that the isovalent substitution of P for As only slightly modified the Fermi surface. So, it is likely that there is competition between $s\pm$-wave and $d$-wave symmetries in these two materials [41].

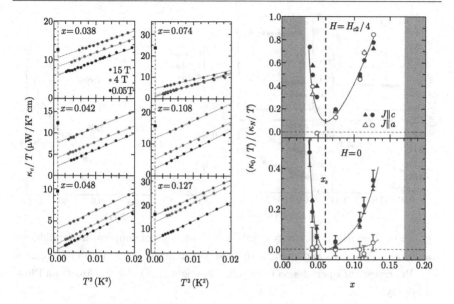

Fig. 12.10   Left panels: Temperature dependencies of the $c$-axis thermal conductivity for Co-Ba122 with different Co concentrations. Right panels: Residual thermal conductivity as a function of Co concentration $x$ at $H = 0$ and $H = H_{c2}/4$ with different thermal current directions. Reprinted with permission from [34]. Copyright (2010) by the American Physical Society.

Tanatar *et al.* measured the very-low-$T$ conductivity of another iron arsenic superconductor, LiFeAs [42], as shown in Fig. 12.11. Obviously, the material had isotropic gap without nodes, and it seems that the symmetry was either $s\pm$-wave or $d$-wave or the community of both. By the way, there is likely a subtle correlation between the isotropic gap and high $T_c$.

Furthermore, the results on thermal conductivity gave a strong evidence for multigap nodeless (at least in $ab$ plane) superconductivity in FeSe$_x$. However, such a gap structure needed further investigations to clarify whether the gap belongs to conventional $s$ wave or unconventional $s\pm$ wave. For Alkali metal doped system $A_x Fe_2 Se_2$ (A = K, Cs), APRES results indicated that conventional $s$-wave symmetry mechanism could describe the system [44], the transport behavior also required further exploration.

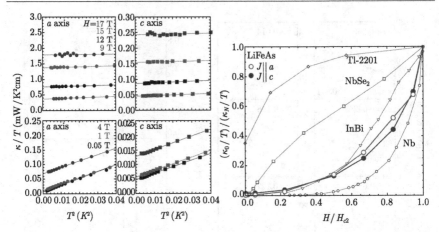

Fig. 12.11   Thermal conductivity of LiFeAs as a function of temperature for different heat current directions and residual thermal conductivity as a function of magnetic field. Reprinted with permission from [42]. Copyright (2011) by the American Physical Society.

Recently, Reid *et al.* [45] studied the very-low-$T$ thermal conductivity of $KFe_2As_2$, and discovered the signature of universal heat transport, which was the first demonstration in the iron-based HTSC. The preceding results told us that $KFe_2As_2$ was $d$-wave nodal superconductor, as shown in Fig. 12.12. Besides, this material exhibited anisotropy obviously. The important finding was that the thermal conductivity of $KFe_2As_2$ was independent of impurity scattering, as shown in Fig. 12.13, the same as the results in YBCO reported by Taillefer *et al.* The nodes in $d$-wave superconductors cannot be removed with impurity scattering, which simply increases the density of QPs. However, impurity scattering will make the gap less anisotropic and eventually remove the nodes in anisotropic $s$-wave superconductors, which also indicated that $KFe_2As_2$ belongs to $d$-wave nodal superconductors.

In summary, due to the more complex band structure and multiple Fermi surfaces, the iron-based superconductors have more complicated gap structures than the cuprate superconductors, which is the main reason for the controversies in the current experimental results.

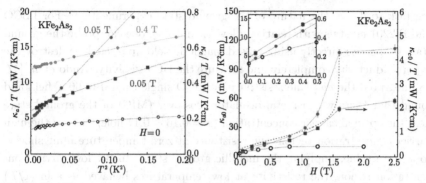

Fig. 12.12   Temperature dependencies of thermal conductivity of KFe$_2$As$_2$ and magnetic field dependencies of the residual thermal conductivity along different directions [45].

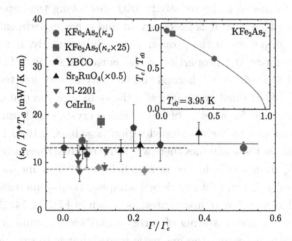

Fig. 12.13   Impurity scattering rate dependence of thermal conductivity. $\Gamma$ represents impurity scattering rate, and $\Gamma \propto \rho_0$, $\rho_0$ is residual resistivity, and $\Gamma_c$ is the critical scattering rate [45].

## 12.4   Very-Low-Temperature Thermal Conductivity and Metal-Insulator Crossover in Cuprate HTSC

The properties of the ground state of HTSC in the normal state have attracted a lot of attention [46]. Since the upper critical field is too high, it is a challenging task to detect the nature of the ground state experimentally. For LSCO and Bi2201 crystals with relatively low $T_c$ and $H_{c2}$, one can probe the ground state properties directly using pulsed magnetic field up to 60 T to destroy the superconductivity completely and measure the DC resistivity. Ando et al. [47, 48] measured the $ab$-plane resistivity of LSCO single crystal in 60 T field and found that there was a metal-insulator crossover (MIC) of the ground state near the optimal carrier concentration ($x = 0.16 \sim 0.17$ hole/Cu). In the high carrier concentration region, the resistivity at zero-temperature limit showed to a finite value, implying the metallic ground state. In the low carrier concentration region, the resistivity at low temperatures behaved as a $\ln(1/T)$ divergence, which is a special insulating ground state and there is no appropriate theoretical explanation at present. For other cuprate superconductors like YBCO and Bi2212, the $H_{c2}$ is too high to achieve in the laboratory, it is not practical to measure the resistivity with destroying the superconductivity completely. Therefore it is necessary to find other experimental methods to detect the properties of the ground state. Subsequently, it is found that one can characterize the properties of the normal state in HTSC from the magnetic-field dependence of thermal conductivity at very low temperatures.

Sun et al. [49] studied systematically the change of thermal conductivity in the magnetic field for a series of LSCO single crystals with various doping from the underdoped to the overdoped region ($x = 0.08, 0.10, 0.14, 0.17$, and $0.22$). Note that in these samples there was finite residual thermal conductivity or the metallic transport behavior in zero field, as already mentioned before. For the optimally doped and overdoped samples, their $\kappa$ increased with the magnetic field at very low temperatures, as shown in Fig. 12.14. At relatively high temperatures, the scattering of vortex on QPs was dominant so that the $\kappa$ decreased with increasing the magnetic field. That is to say, the nature of the ground state can be detected only at very low temperatures. The behavior shown in Fig. 12.14 was firstly discovered by Chiao et al. [50] in optimally doped YBCO, the $\kappa_0/T$ was increased with the applied magnetic field, which resulted from the "Volovik" effect [51]. Therefore, in the metallic phase of superconducting materials, the number of QP excitations can be increased by increasing the magnetic field and lead to the enhancement of $\kappa$, which is a classical behavior of $d$-wave superconductors.

However, in the underdoped LSCO crystals, although the magnetic field could cause an increase in the number of QPs, the low-$T$ thermal conductivity was reduced [49], as shown in Fig. 12.15. It is notable that the "Volovik" effect is the classical behavior of $d$-wave superconductors and is independent of the carrier concentration, which had been supported by low-temperature specific heat results [52]. The reduction of $\kappa$ in the underdoped LSCO with increasing the magnetic field indicated that the QPs were localized. Considering the spin/charge stripe phase in HTSC revealed by neutron scattering [53−55] and STM [56], Sun $et$ $al.$ [49] firstly pointed out that this peculiar field-induced QP localization was closely related to the static spin/charge order caused by the magnetic field in the underdoped HTSC, and this point of view was soon verified by some theoretical works [57, 58]. Clearly, the MIC behavior from the dependence of low-$T$ $\kappa$ on magnetic field is the same as the one by resistivity measurement, both techniques revealed the critical carrier concentration at the optimal doping. This work indicated that the ground state transition from the metallic behavior in the high doping region to the insulating state

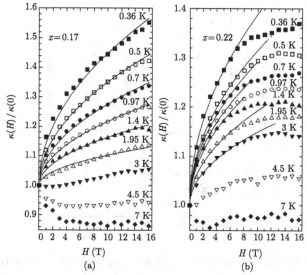

Fig. 12.14   The magnetic-field-induced thermal conductivity of optimally doped and overdoped LSCO single crystals at low temperatures. Reprinted with permission from [49]. Copyright (2003) by the American Physical Society.

in the low doping region can be detected by low-$T$ heat transport, as shown in Fig. 12.16. This conclusion has been further confirmed in the lower temperature thermal conductivity measurements by Hawthorn *et al.* [59].

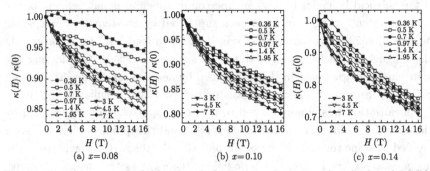

Fig. 12.15   The magnetic-field-induced thermal conductivity of underdoped LSCO single crystals at low temperatures. Reprinted with permission from [49]. Copyright (2003) by the American Physical Society.

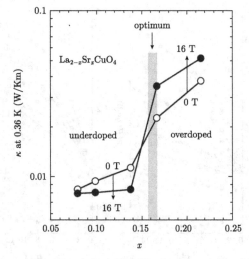

Fig. 12.16   The thermal conductivity of LSCO single crystals with different carrier concentrations at 0.36 K in 0 T and 16 T. The effect of magnetic field on the thermal conductivity demonstrated the crossover of the QP transport properties in the vicinity of the optimal carrier concentration. Reprinted with permission from [49]. Copyright (2003) by the American Physical Society.

For BSLCO system with $T_c = 30$ K, Ando *et al.* [33] proceeded a similar experimental exploration. They measured the thermal conductivity of a series of single crystals with different carrier concentration $(p)$ and then compared the magnetic-field dependences of $\kappa$ at very low temperatures, as shown in Fig. 12.17. The result showed that at 0.36 K, the $\kappa$ of the samples with $p = 0.13$ and 0.15 increased with the magnetic field, while for $p = 0.10$ and 0.11 samples, the $\kappa$ decreased with increasing the magnetic field. Similar to the LSCO system, the dependence of $\kappa$ on the magnetic field at low temperatures indicated that MIC occurred at $p = 0.12$, the same as the previous resistivity measurement under 60 T.

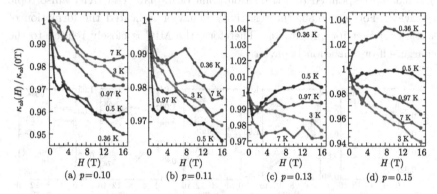

Fig. 12.17   The magnetic-field dependencies of thermal conductivity of BSLCO single crystals with different carrier concentration $(p)$ at low temperatures. Reprinted with permission from [11]. Copyright (2004) by the American Physical Society.

The results of LSCO and BSLCO suggested that the dependence of $\kappa$ on the magnetic field at low temperatures can reflect the MIC of ground state of HTSC, and the magnetic field required is not very high for experiment. It is not unexpected that the transport properties of QPs in the low field should gradually transfer to the normal state behavior with increasing the magnetic field. Although no direct experiments verified whether the transport properties of QPs change in this process, there is indeed reasonable correspondence between those existing phenomena.

So, the low-$T$ thermal conductivity can be used to detect the MIC of the ground state of those HTSC in which the upper critical fields are too high to measure the resistivity, such as YBCO and Bi2212, etc. Sun *et al.* [10]

systematically studied the low-$T$ thermal conductivity of a series of YBCO single crystals with oxygen content from 6.45 to 7.00, as shown in Fig. 12.18. For the optimally doped samples, an early work by Chiao *et al.* [50] had proved that the QP transport was metallic at low temperatures. Results by Sun *et al.* showed that even if the oxygen content was gradually decreased to 6.60, the transport behavior of QP was still metallic at low temperatures, but the samples of oxygen content below 6.50 behaved as localization behavior at low temperatures. Thus, in the YBCO system, the MIC locates at about $y = 6.55$ (the carrier concentration is about 0.07 hole/Cu). The phase diagram of the ground state of YBCO is shown in Fig. 12.19, the author also pointed out that $y = 6.55$ corresponded to a transition point of in-plane resistivity anisotropic behavior. For $y < 6.55$, the resistivity behavior indicated the formation of static charge-stripe phase [60]. Therefore, the MIC is closely related to the charge self-organization behavior.

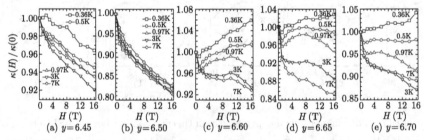

Fig. 12.18  The field dependencies of low-$T$ thermal conductivity for YBCO single crystals with different oxygen contents. Reprinted with permission from [10]. Copyright (2004) by the American Physical Society.

So far, the work by Sun *et al.* [32] is still the only one showing the ground-state properties of YBCO. Recently, Rullier-Albenque *et al.* [61] measured the resistivity of YBCO single crystals with oxygen content of 6.60 in 60 T field and down to about 4 K. It was suggested that the sample exhibited the metallic behavior at low temperatures, which was consistent with the thermal conductivity results by Sun *et al.* But at present there is no resistivity data at lower temperatures and high field in the lower-doped YBCO crystals. Subsequently, the $\mu$SR experiment [62] also demonstrated that the MIC of the ground state in YBCO and LSCO crystals and the peculiar electron localization in the low-doped region are originated from the competition between the field-induced static antiferromagnetism and the superconductivity. On the other hand, neutron scattering [63] indicated that the ground-state MIC in YBCO was associated with the dramatic change of spin-excitation spectrum. The neutron resonance peak and the apparent spin energy gap were likely the fundamental characteristics of the superconductors with metallic ground state, while the superconductors with insulating ground state behaved as gapless spin excitations.

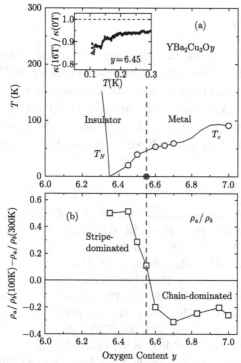

Fig. 12.19  (a) Phase diagram of YBCO, $T_N$ and $T_c$ indicate the antiferromagnetic transition temperature and the superconducting transition temperature, respectively. The red solid dots show the metal-insulator transition point of the ground state, given by the low-$T$ thermal conductivity. (b) The resistivity anisotropy $\rho_a/\rho_b$ at 100 K and 300 K, indicating the formation of charge-stripe phase in the low-doped samples. Reprinted with permission from [10]. Copyright (2004) by the American Physical Society.

In summary, the dependence of thermal conductivity on the magnetic field at low temperatures can probe the nature of the ground state in high field. However, there are still some remaining problems in analyzing the experimental data. In above discussions, the effect of the magnetic field was entirely attributed to the QPs, it is therefore necessary to study further whether the thermal conductivity of phonons is invariant with the magnetic field. Note that some recent papers reported that the phonon conductivity in the parent insulator of cuprates is strongly dependent on the magnetic field [64, 65].

## 12.5 Summary

The very-low-$T$ thermal conductivity is a powerful tool to study the transport behavior of QPs and the superconducting gap symmetry in HTSC. In previous works, many important results, especially the observed finite residual thermal conductivity, directly indicated the unconventional gap symmetry and the existence of nodes. In addition, the magnetic-field-induced dependence of QP thermal conductivity is an efficient technique to detect the nature of the ground state.

[1] R. Berman, *Thermal conduction in solids* (Oxford University Press, 1976).
[2] C. B. Satterthwaite, Phys. Rev. **125**, 873 (1962).
[3] N. E. Hussey, Adv. Phys. **51**, 1685 (2002).
[4] M. J. Graf *et al.*, Phys. Rev. B **53**, 15147 (1996).
[5] A. C. Durst and P. A. Lee, Phys. Rev. B **62**, 1270 (2000).
[6] D. G. Hawthorn *et al.*, Phys. Rev. B **75**, 104518 (2007).
[7] L. Taillefer *et al.*, Phys. Rev. Lett. **79**, 483 (1997).
[8] J. Takeya *et al.*, Phys. Rev. Lett. **88**, 077001 (2002).
[9] M. Sutherland *et al.*, Phys. Rev. B **67**, 174520 (2003).
[10] X. F. Sun, K. Segawa and Y. Ando, Phys. Rev. Lett. **93**, 107001 (2004).
[11] Y. Ando *et al.*, Phys. Rev.Lett. **92**, 247004 (2004).
[12] N. E. Hussey *et al.*, Phys. Rev. Lett. **85**, 4140 (2000).
[13] X. F. Sun *et al.*, Phys. Rev. Lett. **96**, 017008 (2006).
[14] S. H. Pan *et al.*, Nature **403**, 746 (2000).
[15] S. H. Pan *et al.*, Nature **413**, 282 (2001).
[16] C. Howald, P. Fournier and A. Kapitulnik, Phys. Rev. B **64**, 100504(R) (2001).
[17] K. M. Lang *et al.*, Nature **415**, 412 (2002).

[18]  K. McElroy et al., Science **309**, 1048 (2005).

[19]  K. K. Gomez et al., Nature **447**, 569 (2007).

[20]  A. N. Pasupathy et al., Science **320**, 196 (2008).

[21]  W. A. Atkinson and P. J. Hirschfeld, Phys. Rev. Lett. **88**, 187003 (2002).

[22]  B. M. Andersen and P. J. Hirschfeld, Phys. Rev. Lett. **100**, 257003 (2008).

[23]  H. Ding et al., Europhys. Lett. **83**, 47001 (2008).

[24]  R. T. Gordon et al., Phys. Rev. Lett. **102**, 127004 (2009).

[25]  R. T. Gordon et al., Phys. Rev. B **79**, 100506(R) (2009).

[26]  K. Terashima et al., Proc. Natl. Acad. Sci. **106**, 7330 (2009).

[27]  L. Zhao et al., Chin. Phys. Lett. **25**, 4402 (2008).

[28]  K. Nakayama et al., Europhys. Lett. **85**, 67002 (2009).

[29]  M. A. Tanatar et al., Phys. Rev. Lett. **104**, 067002 (2010)

[30]  C. Proust et al., Phys. Rev. Lett. **89**, 147003 (2002).

[31]  F. Wang et al., Phys. Rev. Lett. **102**, 047005 (2009).

[32]  V. Mishra et al., Phys. Rev. B **79**, 094512 (2009).

[33]  A.V. Chubukov et al., Phys. Rev. B **80**, 140515(R) (2009).

[34]  J. P. Reid et al., Phys. Rev. B **82**, 064501 (2010).

[35]  L. Ding et al., New J. Phys. **11**, 093018 (2009).

[36]  M. A. Tanatar et al., Phys. Rev. Lett. **104**, 067002 (2010).

[37]  J. K. Dong et al., Phys. Rev. B **81**, 094520 (2010).

[38]  X. G. Luo et al., Phys. Rev. B **80**, 140503(R) (2009).

[39]  P. Richard et al., Phys. Rev. Lett. **102**, 047003 (2009).

[40]  T. Sato et al., Phys. Rev. Lett. **103**, 047002 (2009).

[41]  J. K. Dong et al., Phys. Rev. Lett. **104**, 087005 (2010).

[42]  M. A. Tanatar et al., Phys. Rev. B **84**, 054507 (2011).

[43]  J. K. Dong et al., Phys. Rev. B **80**, 024518 (2009).

[44]  Y. Zhang et al., Nature Mater. **10**, 273 (2011).

[45]  J. P. Reid et al., arXiv: cond-mat/ 1201.3376.

[46]  M. Imada, A. Fujimori and Y. Tokura, Rev. Mod. Phys. **70**, 1039 (1998).

[47]  Y. Ando et al., Phys. Rev. Lett. **75**, 4662 (1995).

[48]  G. S. Boebinger et al., Phys. Rev. Lett. **77**, 5417 (1996).

[49]  X. F. Sun et al., Phys. Rev. Lett. **90**, 117004 (2003).

[50]  M. Chiao et al., Phys. Rev. Lett. **82**, 2943 (1999).

[51]  G. E. Volovik, JETP Lett. **58**, 469 (1993).

[52]  S. J. Chen et al., Phys. Rev. B **58**, R14753 (1998).

[53]  B. Lake et al., Science **291**, 1759 (2001).

[54]  B. Lake et al., Nature **415**, 299 (2002).

[55]  B. Khaykovich et al., Phys. Rev. B **66**, 014528 (2002).

[56]  J. E. Hoffman et al., Science **295**, 466 (2002).

[57]  V. P.Gusynin and V. A. Miransky, Eur. Phys. J. B **37**, 363 (2004).

[58] M. Takigawa, M. Ichioka, and K. Machida, Physica C **404**, 375 (2004).

[59] D. G. Hawthorn *et al.*, Phys. Rev. Lett. **90**, 197004 (2003).

[60] Y. Ando *et al.*, Phys. Rev. Lett. **88**, 137005 (2002).

[61] F. Rullier-Albenque *et al.*, Phys. Rev. Lett. **99**, 027003 (2007).

[62] J. E. Sonier *et al.*, Phys. Rev. B **76**, 064522 (2007).

[63] S. L. Li *et al.*, Phys. Rev. B **77**, 014523 (2008).

[64] X. F. Sun *et al.*, Phys. Rev. B **72**, 104501 (2005).

[65] Z. Y. Zhao *et al.*, Phys. Rev. B **83**, 174518 (2011).

# 13
# Off-Diagonal Long Range Order in Emergent Quantum Phases

G. Su

*Theoretical Condensed Matter Physics and Computational Materials Physics Laboratory, School of Physics, University of Chinese Academy of Sciences, P. O. Box 4588, Beijing 100049, China*

Superconductors and superfluids are quantum phases characterized by the existence of off-diagonal long range order (ODLRO), and both share some similarities between physical properties. By briefly retrospecting the historic developments of superconductivity, Bose-Einstein condensation and superfluidity, we discuss the property of ODLRO and its consequences as well as the relevance to other emergent phenomena in physics. The ODLRO in several quantum systems is also reviewed, and the concept of superstate is introduced. Finally, a prospect for high temperature superconductivity is outlined. This paper is dedicated to the 100 anniversary of the discovery of superconductivity.

## 13.1 Superconductivity

One hundred years ago, on 8 April 1911, the phenomenon of superconductivity in solid mercury (Hg) was discovered by a Dutch physicist Heike Kamerlingh Onnes [1]. During this century-long period, since superconductors were found to have a plenty of practical applications in various fields, people have been full of enthusiasms to explore superconductors with high transition temperatures or even seeking for room-temperature superconductors, and rich achievements have been made. Meanwhile, the sustaining investigations on superconducting (SC) materials and superconductivity led to the establishment of some new concepts and theories, and new phenomena were also disclosed, boosting remarkably the development of the condensed matter physics and related areas.

In fact, only two years after Kamerlingh Onnes discovered the superconductivity in solid Hg at 4.2 K, people found that metallic lead (Pb) also exhibited superconductivity below 7 K. At the same year (1913), Kamerlingh Onnes was awarded a Nobel prize in physics. After then, the phenomenon of superconductivity was successively observed in other elements. At present, it is known that in the periodic table of elements, 30 elements superconduct below certain critical temperatures, and 23 elements superconduct under high pressures. Nonetheless, the elemental superconductors have low SC transition temperatures. At ambient pressure, the SC transition temperature ($T_c$) of niobium (Nb) bears the highest record, namely 9.3 K; under high pressure, the highest $T_c$ is from calcium (Ca), $T_c = 25$ K at pressure 161 GPa. $T_c$ is 19.6 K for scandium (Sc) at 106 GPa, 19.5 K for yttrium (Y) at 115 GPa, 17.3 K for sulphur (S) at 190 GPa. Because of their low SC transition temperatures, the extensive industrial applications of these SC elements are quite limited. At that time, superconductors with higher $T_c$ usually come from alloys and compounds. Till 1979, the typical materials of such kind of superconductors included NbN, $V_3Si$, $Nb_3Sn$, $Nb_3Ge$, etc., where $Nb_3Ge$ film has the highest $T_c$ of 23.2 K among them.

During that period, the SC theories have also made some milestoned breakthroughs. In 1933, Meissner and Ochsenfeld uncovered the effect of complete diamagnetism of superconductors (now known as Meissner effect) [2]. In 1935, F. London and H. London proposed the electromagnetic theory of superconductivity, phenomenologically explaining the two fundamental properties, say, zero resistance and complete diamagnetism, of superconductors [3]. In 1950, Ginzburg and Landau constructed, based on Landau's theory of the second-order phase transition, a phenomenological theory of superconductivity (now known as Ginzburg-Landau phenomenological theory [4]). Starting from this theory, in 1957, Abrikosov established the concept of type-II superconductors, and predicted the existence of magnetic flux vortex lattice [5]. Abrikosov and Ginzburg were thus awarded the Nobel prize in physics in 2003. Also in 1950, two groups observed, independently, the isotope effect in Hg superconductor [6], unveiling that the electron-phonon interaction could be related to the origin of superconductivity. In 1957, Bardeen, Cooper and Schrieffer published the celebrated microscopic theory of superconductivity (now known as BCS theory [7]), proposed that superconductivity comes from the condensation of electron pairs (now known as Cooper pairs) that are formed in momentum space by exchanging the virtual phonons between two electrons with opposite spin and momentum. BCS theory can explain almost all properties and phenomena observed previously in superconductors in experiments, and gave nontrivial predictions that were also confirmed in experiments afterward. Bardeen, Cooper and Schrieffer shared the Nobel prize in 1972. In 1960, Eliashberg proposed the so-called BCS strong-coupling theory of superconductivity [8]. In 1962, a 22-year-old British student, Brian D. Josephson, predicted theoretically the tunneling effect of Cooper pairs [9], namely, a direct supercurrent can pass through a junction that consists of two superconductors separated by a thin insulating layer at zero bias voltage, and an alternate supercurrent can flow through such a junction at a direct-current bias voltage, which was confirmed quickly in experiments. This effect is now known as Josephson effect, which is nowadays widely applied in the field of SC quantum interference devices. Josephson was thus awarded Nobel prize in 1973 for predicting such an effect.

Now the superconductors that can be well explained within the framework of BCS theory are usually referred to as the conventional superconductors or low temperature superconductors, and those beyond the BCS theory are usually called unconventional superconductors. Over decades, it was believed that the SC transition temperature of superconductors would not be higher than 30 K based on the BCS theory. Such an argument was undermined in 1986, when Bednorz and Müller diclosed that the perovskite cuprate oxide ceramic compound LaBaCuO could show superconductivity below temperature 35 K [10], which was quickly confirmed worldwide. Next year, Bednorz and Müller were awarded the Nobel prize. Afterwards, a vast number of cuprate oxide superconductors were found, where, among others, typical examples include YBaCuO ($T_c = 90$ K), BiSrCaCuO ($T_c = 110$ K), TlBaCaCuO ($T_c = 125$ K), HgBaCaCuO ($T_c = 135$ K), etc., and under high pressure 30 GPa, $T_c$ of $HgBa_2Ca_2Cu_3O_8$ can be as high as 164 K, which is the highest value of SC transition temperatures of superconductors confirmed experimentally up to date. Other superconductors, such as heavy fermion superconductors ($CeCu_2Si_2$,

$UBe_{13}$, $UPt_{13}$, $UPd_2Al_{13}$, $CeCoIn_5$, $PuCoGa_5$, etc.), organic superconductors ($K_3C_{60}$, $Cs_3C_{60}$, $RbCs_2C_{60}$, $CaC_6$, $YbC_6$, $NaC_2$, $K_{3.3}$ Picene, $K_3$ Phenanthrene, B doped diamond, etc.), $MgB_2$, $Na_xCoO_2 \cdot yH_2O$, $SiH_4$ under high pressure, and so on, have also received much attention.

In 2008, Hosono *et al.* discovered that superconductivity can appear in the compound $LaO_{1-x}F_xFeAs$ with $T_c = 26$ K [11]. Just in a few months, the other iron-based superconductors with higher $T_c$, including 1111-, 122-, 111-, and 11-phase, such as $SmFeAsO_{0.85}$, $Sr_{0.5}Sm_{0.5}FeAsF$, $BaFe_2As_2$, $LiFeAs$, $NaFeAs$, $FeSe$, $K_{0.8}Fe_2Se_2$, etc., were found, with the highest $T_c = 56$ K, booming a new round of tide of exploring high temperature superconductors. Very recently, new layered boron nitrogen superconductors, like $LaNiBN$ ($T_c \simeq 4.1$ K), $CaNiBN$ ($T_c \simeq 2.2$ K), $LaPtBN$ ($T_c \simeq 6.7$ K), etc., have been synthesized [12].

One of central missions for studying superconductivity is to seek for or to discover room temperature superconductors. In 1968, Ashcroft predicted, theoretically, the metallic hydrogen could show room temperature superconductivity under extremely high pressure (500 GPa) [13], but it is not yet confirmed with a successful experiment. The other examples reported gloomy hints of room temperature superconductors like $PdH_x$ ($x > 1$) [14], and $(Tl_5Pb_2)Ba_2MgCu_{10}O_{17+}$ [15]. But they were also not confirmed by other people. Nevertheless, great efforts in experiments and theories are still being paid on exploring new and unconventional superconductors (see, e.g. Refs. [16, 17], and other references for review).

## 13.2　Bose-Einstein Condensation and Superfluidity in $^4$He

In 1924, S. N. Bose, a 30-year-old associate professor in physics at Dhaka University, India, wrote a short article with a new idea on Planck's law of black-body radiation and the hypothesis of light quanta, where he did not invoke the traditional classical law in physics, but instead, he adopted directly a statistical method for counting the states of identical particles, and obtained the result in agreement with experiments. When Bose submitted his short paper to various journals in physics for publication, he was rejected by the editors. With frustration, he sent this paper to Einstein, who realized quickly the importance of Bose's method. By translating this paper from English to

German himself, Einstein recommended it with his comments for publication in Zeitschrift für Physik [18]. Subsequently, Einstein generalized Bose's statistical method on photons to massive identical particles with spin integers (bosons), and proposed the celebrated Bose-Einstein statistics [19]. Einstein further disclosed that, when the identical bosons are cooled down to a very low temperature, these identical bosons in momentum space can condense into a macroscopically occupied quantum state with momentum zero, forming a new state of matter. This is nothing but the famous Bose-Einstein condensation (BEC).

It is well-known that Bose and Einstein's theory was initially proposed for noninteracting Bose systems. For an isotropic system composed of $N$ noninteracting bosons with dispersion relation given by $\epsilon_p = \gamma |p|^\alpha$, in the thermodynamic limit ($N \to \infty$, $V \to \infty$ but the density $n = N/V$ fixed) a phase transition can be found at a critical temperature $T_c$ [20],

$$T_c = \frac{[2^{1-(1/d)}\pi^{1/2}\hbar]^\alpha \gamma}{k_B} \left[ \frac{\Gamma(d/2)n\alpha}{\Gamma(d/\alpha)\zeta(d/\alpha)} \right]^{\alpha/d}, \tag{13.1}$$

where $\alpha$ and $\gamma$ are positive constants, $d$ is the spatial dimensionality, $\Gamma(x)$ is the gamma function and $\zeta(x) = \sum_{l=0}^{\infty} l^{-x} (l > 0)$ is the Riemann Zeta function. When $\alpha = 2$, $\gamma = 1/(2m)$, $d = 3$, $T_c \approx 3.31\hbar^2 n^{2/3}/(mk_B)$. The condensed fraction of bosons $(n_0)$ is given by $n_0 = n[1 - (T/T_c)^{d/\alpha}]$. The Bernouli equation for the system satisfies $PV = (\alpha/d)E$ with $P$ the pressure and $E$ the internal energy [21]. The specific heat can be obtained for $T < T_c$ by

$$c_v = \frac{[(d/\alpha) + 1]k_B}{2^{d-1}\pi^{d/2}\hbar^d \alpha} \frac{\Gamma[(d/\alpha) + 1]\zeta[(d/\alpha) + 1]}{\Gamma(d/2)} \left( \frac{k_B T}{\gamma} \right)^{d/\alpha}, \tag{13.2}$$

and when $T > T_c$ the effect of the temperature-dependence of the chemical potential $\mu = \mu(T)$ must be considered at a fixed density, and at high temperatures $c_v \to (d/\alpha)k_B n$. The phase transition is of first-order for an ideal Bose gas, but it is not for an interacting Bose system, e.g. the $\lambda$-transition of liquid helium is of second-order.

The BEC has not received much attention for quite long time since its occurrence, even including Einstein himself. The main reason is that it is hard to realize the ideal Bose system in laboratory. The situation dramatically changed after the discovery of superfluidity in liquid $^4$He, which led to

London's proposing that the transition from He I to He II phase is an analog of BEC, i.e., the superfluidity in He II is characterized by the macroscopic occupation of a single quantum state [22], although the liquid He II is in fact a strongly interacting Bose system. Following London's suggestion, Tisza [23], while Landau simultaneously [24], developed phenomenologically a two-fluid model, in which the superfluid $^4$He is regarded as a combination of superfluid and normal fluid components. The superfluid component is a single condensed quantum state, carrying no entropy, and thereby resulting in the assumption that the superfluid flow is irrotational. A few years later, Onsager [25] suggested that quantized vortex lines, or the quantization of vorticity according to Feynman independently [26], must exist in $^4$He, which was experimentally verified later [27]. Since then, the fact that the irrotational flow and the quantization of circulation (vortices) are two fundamental characters of superfluid $^4$He was well established. Landau was awarded the Nobel prize in 1962 for his essential contribution to the theory of superfluidity of $^4$He. Later, superfluidity was also observed in $^3$He at low temperature. Because $^3$He atoms are fermions, superfluidity comes from the condensation of $^3$He atom pairs, similar to the Cooper pairs of electrons in BCS theory. The experimental and theoretical works on the superfluidity in $^3$He were awarded Nobel physics prizes in 1996 and 2003, respectively.

After seventy years of theoretical proposal of the BEC, on 5 June 1995, Eric Cornell, Carl Wieman and co-workers [28] of Colorado University, USA, combined laser cooling and magnetic evaporating cooling techniques, and cooled the dilute vapor composed of two thousand $^{87}$Rb atoms down to extremely low temperature 187 nK, realized the BEC of cold atoms in laboratory. A few months later, Wolfgang Ketterle and co-workers [29] of Massachusetts Institute of Technology, USA, generated the BEC of a half million cold $^{23}$Na atoms, and observed the quantum interference between two condensates of $^{23}$Na atoms. Cornell, Wieman and Ketterle thus shared the Nobel physics prize in 2001. Henceforth, many Bose-Einstein condensates of cold atoms like $^7$Li, $^{39}$K, $^{41}$K, $^{85}$Rb, $^{133}$Cs, $^{52}$Cr, $^{40}$Ca, $^{84}$Sr, $^{86}$Sr, $^{88}$Sr, $^{174}$Yb, etc. have been produced.

Besides, by means of the BEC states of cold atoms, some basic phenomena in physics such as atomic laser, matter wave solitons, superfluidity and quantum vortex, optical lattice, superfluidity-Mott insulator transition, BCS-BEC crossover, Josephson effect, slowing down the velocity of light pulses with electromagnetically induced transparency technique, simulating black holes, simulating strongly correlated fermions and bosons, and so on, can be studied in laboratory. By placing the cold atoms on the optical lattice, one can simulate the motion of electrons on a periodic lattice in solids, and could also generate n- and p-type doped semiconducting states. A new discipline — atomtronics [30] — is emerging. Certainly, BEC can also occur in the systems of bosonic quasiparticles in solids such as phonons, magnons, excitons, etc.

## 13.3 Off-Diagonal Long Range Order

As mentioned above, the BEC was first proposed for noninteracting Bose systems. Is it available for an interacting Bose system? Bogoliubov [31] answered this question by finding that the BEC was not changed so much by interactions in weakly interacting dilute Bose systems. Several years later, Penrose and Onsager [32] considered this nontrivial question by generalizing the mathematical description of BEC to interacting Bose systems, and manifested that BEC appears when a macroscopically large eigenvalue of the one-particle reduced density matrix exists, and the liquid helium II was shown to possess such a property. Later, the concept was successfully extended to superconductivity in interacting fermion systems by C. N. Yang [33] who first coined it as the off-diagonal long range order (ODLRO) and believed that superconductivity is characterized by the ODLRO of the two-particle reduced density matrix. Even though, this concept was not immediately accepted. After several years people gradually realized that this important description is deeply related to the spontaneous symmetry breaking by Goldstone [34] and the phase coherence [35]. It is now well accepted that both superconductivity and superfluidity are quantum phases characterized by the existence of ODLRO. In the following, let us address briefly the concept of ODLRO and its physical consequences.

The one-particle reduced density matrix $\rho_1(\boldsymbol{r}, \boldsymbol{r}')$ is defined as

$$\rho_1(\boldsymbol{r}, \boldsymbol{r}') = \langle a_{\boldsymbol{r}}^{\dagger} a_{\boldsymbol{r}'} \rangle = \frac{1}{V} \sum_{\boldsymbol{p}, \boldsymbol{q}} e^{-i(\boldsymbol{p}\cdot\boldsymbol{r} - \boldsymbol{q}\cdot\boldsymbol{r}')} \langle a_{\boldsymbol{p}}^{\dagger} a_{\boldsymbol{q}} \rangle, \tag{13.3}$$

where $a_{\boldsymbol{r}}^{\dagger}$ is the creation operator of a boson at position $\boldsymbol{r}$, $a_{\boldsymbol{p}}^{\dagger}$ is the Fourier transform of $a_{\boldsymbol{r}}^{\dagger}$, and $\langle \cdots \rangle$ denotes the thermal average. Eq. (13.3) can be rewritten as

$$\rho_1(\boldsymbol{r},\boldsymbol{r}') = \frac{1}{Z} \int \cdots \int \frac{d\boldsymbol{r}_2 \cdots d\boldsymbol{r}_N}{(N-1)!} \sum_n e^{-E_n/k_B T}$$

$$\psi_n(\boldsymbol{r},\boldsymbol{r}_2,\cdots,\boldsymbol{r}_N)\psi_n^*(\boldsymbol{r}',\boldsymbol{r}_2,\cdots,\boldsymbol{r}_N), \qquad (13.4)$$

where $Z = \sum_n e^{-E_n/k_B T}$ is the partition function, $\psi_n(\boldsymbol{r}_1,\boldsymbol{r}_2,\cdots,\boldsymbol{r}_N)$ is $N$-particle wave function of the Hamiltonian of the system, and $E_n$ is the corresponding eigenvalue. The normalization of $\rho_1(\boldsymbol{r},\boldsymbol{r}')$ satisfies $\mathrm{Tr}\rho_1 = \int d\boldsymbol{r}\rho_1(\boldsymbol{r},\boldsymbol{r}) = N$, and $\rho_1(\boldsymbol{r},\boldsymbol{r}')$ is single-valued.

Owing to the translational invariance, the total momenta $\wp = \sum_{\boldsymbol{p}} \boldsymbol{p} a_{\boldsymbol{p}}^\dagger a_{\boldsymbol{p}}$ commutes with the Hamiltonian of the system, leading to $\langle [\wp, a_{\boldsymbol{p}}^\dagger a_{\boldsymbol{q}}] \rangle = 0$, which yields $\langle a_{\boldsymbol{p}}^\dagger a_{\boldsymbol{q}} \rangle = \langle n_{\boldsymbol{p}} \rangle \delta_{\boldsymbol{pq}}$ with $n_{\boldsymbol{p}} = a_{\boldsymbol{p}}^\dagger a_{\boldsymbol{p}}$. Consequently, we have

$$\rho_1(\boldsymbol{r},\boldsymbol{r}') = \frac{1}{V} \sum_{\boldsymbol{p}} e^{-i\boldsymbol{p}\cdot(\boldsymbol{r}-\boldsymbol{r}')} \langle n_{\boldsymbol{p}} \rangle$$

$$= \frac{\langle n_0 \rangle}{V} + \frac{1}{V} \sum_{\boldsymbol{p}(\neq 0)} e^{-i\boldsymbol{p}\cdot(\boldsymbol{r}-\boldsymbol{r}')} \langle n_{\boldsymbol{p}} \rangle. \qquad (13.5)$$

The second term of the right-hand side of Eq.(13.5) tends to zero as the off-diagonal long rang (ODLR) limit $|\boldsymbol{r}-\boldsymbol{r}'| \to \infty$. If $\langle n_0 \rangle$ is macroscopically large, i.e. $\langle n_0 \rangle = \alpha_0 N$ with $\alpha_0$ a finite fraction on the order of unity, and therefore $\rho_1(\boldsymbol{r},\boldsymbol{r}') \to \alpha_0(N/V) \neq 0$ as $|\boldsymbol{r}-\boldsymbol{r}'| \to \infty$, then the Bose system is said to possess the ODLRO in $\rho_1$. For an ideal Bose gas as well as in liquid $^4$He, it has been shown that such an order exists below a certain temperature. If the condensation occurs in a state with $\boldsymbol{p} \neq \boldsymbol{0}$, it can be shown that

$$\rho_1(\boldsymbol{r},\boldsymbol{r}') = \langle \varphi_{\boldsymbol{p}} | a_{\boldsymbol{r}}^\dagger a_{\boldsymbol{r}'} | \varphi_{\boldsymbol{p}} \rangle \to \alpha_0 \left(\frac{N}{V}\right) e^{-i\boldsymbol{p}\cdot(\boldsymbol{r}-\boldsymbol{r}')}, \text{ as } |\boldsymbol{r}-\boldsymbol{r}'| \to \infty. \quad (13.6)$$

For example, when the wave function bears the form of $|\varphi_{\boldsymbol{p}}\rangle = (1/\sqrt{N_0!})(a_{\boldsymbol{p}}^\dagger)^{N_0}|0\rangle$ with $N_0$ the number of condensed particles, a simple calculation gives $\rho_1(\boldsymbol{r},\boldsymbol{r}') = \langle \varphi_{\boldsymbol{p}} | a_{\boldsymbol{r}}^\dagger a_{\boldsymbol{r}'} | \varphi_{\boldsymbol{p}} \rangle = (N_0/V) \exp[-i\boldsymbol{p} \cdot (\boldsymbol{r}-\boldsymbol{r}')]$. As emphasized by Penrose and Onsager [32], the state $|\varphi_{\boldsymbol{p}}\rangle$, with the largest eigenvalue being extensive, is not necessarily the lowest energy level. Actually, the criterion of Eqs. (13.5) and (13.6) implies the cluster separation,

$$\rho_1(\boldsymbol{r}, \boldsymbol{r}') = \langle a_{\boldsymbol{r}}^\dagger a_{\boldsymbol{r}'} \rangle \to \langle a_{\boldsymbol{r}}^\dagger \rangle \langle a_{\boldsymbol{r}'} \rangle, \text{ as } |\boldsymbol{r} - \boldsymbol{r}'| \to \infty. \tag{13.7}$$

The function $\langle a_{\boldsymbol{r}}^\dagger \rangle$ is a complex function, with definite amplitude and phase, which can be identified as the condensate wave function, or the order parameter of the BEC system.

On the other hand, we can re-express $\rho_1(\boldsymbol{r}, \boldsymbol{r}')$ in the spectral form of

$$\rho_1(\boldsymbol{r}, \boldsymbol{r}') = \sum_\nu \lambda_\nu \Phi_\nu(\boldsymbol{r}) \Phi_\nu^*(\boldsymbol{r}'), \tag{13.8}$$

where $\Phi_\nu$ is the eigenfunctions of $\rho_1$, with eigenvalues $\lambda_\nu$. The existence of ODLRO in $\rho_1$ implies the spectral resolution such that

$$\rho_1(\boldsymbol{r}, \boldsymbol{r}') = \lambda_0 \Phi_0(\boldsymbol{r}) \Phi_0^*(\boldsymbol{r}') + \rho_1'(\boldsymbol{r}, \boldsymbol{r}'), \tag{13.9}$$

where $\lambda_0$ is the largest eigenvalue of $\rho_1$, $\Phi_0(\boldsymbol{r})$ is the corresponding eigenfunction (or the condensate wave-function), and $\rho_1'(\boldsymbol{r}, \boldsymbol{r}')$ denotes other terms. In the thermodynamic limit, because $\lambda_0 \nrightarrow 0$ and $\rho_1'(\boldsymbol{r}, \boldsymbol{r}') \to 0$ as $|\boldsymbol{r} - \boldsymbol{r}'| \to \infty$, $\rho_1(\boldsymbol{r}, \boldsymbol{r}')$ is nonzero. It should be mentioned that the definitions in Eqs. (13.5), (13.7) and (13.9) are equivalent actually.

The two-particle reduced density matrix $\rho_2(\boldsymbol{r}_1, \boldsymbol{r}_2; \boldsymbol{r}_1', \boldsymbol{r}_2')$ is defined in the similar way,

$$\rho_2(\boldsymbol{r}_1, \boldsymbol{r}_2; \boldsymbol{r}_1', \boldsymbol{r}_2') = \langle a_{\boldsymbol{r}_1}^\dagger a_{\boldsymbol{r}_2}^\dagger a_{\boldsymbol{r}_1'} a_{\boldsymbol{r}_2'} \rangle, \tag{13.10}$$

which can also be represented as

$$\rho_2(\boldsymbol{r}_1, \boldsymbol{r}_2; \boldsymbol{r}_1', \boldsymbol{r}_2') = \frac{1}{Z} \int \cdots \int \frac{d\boldsymbol{r}_3 \cdots d\boldsymbol{r}_N}{(N-2)!} \sum_n e^{-E_n/k_B T}$$
$$\times \psi_n(\boldsymbol{r}_1, \boldsymbol{r}_2, \boldsymbol{r}_3, \cdots, \boldsymbol{r}_N) \psi_n^*(\boldsymbol{r}_1', \boldsymbol{r}_2', \boldsymbol{r}_3, \cdots, \boldsymbol{r}_N). \tag{13.11}$$

The normalization of $\rho_2(\boldsymbol{r}_1, \boldsymbol{r}_2; \boldsymbol{r}_1', \boldsymbol{r}_2')$ satisfies

$$\mathrm{Tr}\rho_2 = \int \int \rho_2(\boldsymbol{r}_1, \boldsymbol{r}_2; \boldsymbol{r}_1, \boldsymbol{r}_2) d\boldsymbol{r}_1 d\boldsymbol{r}_2 = N(N-1). \tag{13.12}$$

In the spectral form, we have

$$\rho_2(\boldsymbol{r}_1, \boldsymbol{r}_2; \boldsymbol{r}_1', \boldsymbol{r}_2') = \sum_\nu \alpha_\nu \Phi_\nu(\boldsymbol{r}_1, \boldsymbol{r}_2) \Phi_\nu^*(\boldsymbol{r}_1', \boldsymbol{r}_2'), \tag{13.13}$$

where $\alpha_\nu$ is the eigenvalue of $\rho_2$, and $\Phi_\nu(r_1, r_2)$ is the corresponding normalized eigenfunction. The existence of ODLRO in the two-particle reduced density matrix $\rho_2$ implies that in the ODLR limit ($|r_i - r'_j| \to \infty$, $i, j = 1, 2$, but $|r_1 - r_2|$ and $|r'_1 - r'_2|$ finite), the following spectral resolution exists,

$$\rho_2(r_1, r_2; r'_1, r'_2) \to \lambda_0 \Phi_0(r_1, r_2) \Phi_0^*(r'_1, r'_2), \qquad (13.14)$$

where $\lambda_0$ is the largest eigenvalue of $\rho_2$, and $\Phi_0$ is the corresponding eigenfunction.

For the Bose and Fermi systems with nonzero spin, the single- and two-particle reduced density matrices depend in general on spin $\sigma$. Their spectral resolutions can be written as

$$\rho_1(r, \sigma; r', \sigma') = \lambda_0 \Phi_0(r,\sigma) \Phi_0^*(r', \sigma') + \rho'_1(r, \sigma; r', \sigma'), \quad (13.15)$$

$$\begin{aligned} \rho_2(r_1, \sigma_1, r_2, \sigma_2; r'_1, \sigma'_1, r'_2, \sigma'_2) = {} & \lambda_0 \Phi_0(r_1, \sigma_1, r_2, \sigma_2) \Phi_0^*(r'_1, \sigma'_1, r'_2, \sigma'_2) \\ & + \rho'_2(r_1, \sigma_1, r_2, \sigma_2; r'_1, \sigma'_1, r'_2, \sigma'_2), \end{aligned} \qquad (13.16)$$

where in the thermodynamic limit as $|r - r'| \to \infty$, $\rho'_1 \to 0$; $|(r_1, r_2) - (r'_1, r'_2)| \to 0$, $\rho'_2 \to 0$. Obviously, the wave function $\Phi_0(r,\sigma)$ or $\Phi_0(r_1, \sigma_1, r_2, \sigma_2)$ of the condensate is a spinor. For a Bose system, in the presence of a magnetic field, the time reversal symmetry can be destroyed, and the spatial change of magnetic field could result in a local spin gauge symmetry, inducing vortices [36].

It has been proven [33] that, for a Bose system, $\rho_1$ can possess the ODLRO, and $\rho_2$ can also. Therefore, to judge whether or not the BEC can take place in a Bose system, one only needs to study if the ODLRO exists in $\rho_1$. For an interacting Fermi system, in the thermodynamic limit, $\rho_1 = 0$, no ODLRO exists there, but the ODLRO may exist in $\rho_2$. Thus, to judge if the superconducting condensation can appear in a Fermi system, one needs to explore the two-particle reduced density matrix $\rho_2$. For $N$-fermion system in $M$ states, the largest eigenvalue of $\rho_2$ satisfies $\alpha_0 \leqslant N(M - N + 2)/M$ [33].

It is common knowledge that two fundamental characters of a superconductor are Meissner effect and flux quantization, while two fundamental characters of a superfluid are irrotational flow and quantization of vortices. An interesting but crucial question arises: is there any relationship between the existence of ODLRO and these fundamental properties of superconductor and superfluid? The answer to this question is yes. By a rigorous proof mathematically, it is disclosed that for an interacting Fermi system, the existence of ODLRO in the two-particle reduced density matrix implies the Meissner effect and flux quantization [37,38], and uncovered that the phenomena of Meissner effect and flux quantization are closely related, and the underlying mechanism behind them is the phase coherence of the wave function of the SC condensate. For an interacting Bose system, we proposed and proved a theorem [39] that, the existence of ODLRO in the single-particle reduced density matrix implies the irrotational flow and quantization of vortices, and the phase coherence of the wave function of the superfluid condensate is the origin between both. It is also proved that for a Bose condensate, the density has the translational invariance. In this way, by means of the local gauge invariant principle we have established an underlying connection between the ODLRO and superconductors and superfluids, strengthened further that superconductors and superfluids are quantum phases characterized by the existence of ODLRO. This also gives a strong support that the ODLRO can be taken as a criterion for superconductivity and superfluidity.

For an ideal Bose gas, there is the ODLRO in $\rho_1$. According to the above theorem, the Bose-Einstein condensate should possess the properties of irrotational flow and quantization of vortices, which is also able to be verified by the BEC experiments of cold atoms. Therefore, the ideal Bose gas is a superfluid. It does not possess the superfluidity, however. In accordance with Landau's criterion for superfluidity, the critical velocity of an ideal Bose gas is zero, suggesting that it is unstable against any motion of the system. At finite temperature, the ODLRO appears below the critical temperature, the onset of which is accompanied with the singularities of the specific heat and compressibility.

The existence of ODLRO is associated with the gauge symmetry breaking [39]. In the Bose condensate, the continuity equation becomes $\partial\phi_0/\partial t + \boldsymbol{v}_s \cdot \nabla\phi_0 + (1/2)\phi_0 \nabla \cdot \boldsymbol{v}_s = 0$. After integrating it, we have

$$\frac{\partial}{\partial t} \int d\boldsymbol{r}\phi_0(\boldsymbol{r},t) = \frac{1}{2} \int d\boldsymbol{r}\phi_0(\boldsymbol{r},t)\nabla \cdot \boldsymbol{v}_s(\boldsymbol{r})$$
$$- \int d\boldsymbol{r}\nabla \cdot [\boldsymbol{v}_s(\boldsymbol{r})\phi_0(\boldsymbol{r},t)].$$

In view of Gaussian theorem, the second term of the right-hand side of above equation should be a surface integral, and is vanishing in the present case. For a steady flow, $\nabla \cdot \boldsymbol{v}_s(\boldsymbol{r}) = 0$, the above equation tells us that $\int d\boldsymbol{r}\phi_0(\boldsymbol{r}, t)$ is independent of time. If we define the spatial Fourier transform of $\phi_0(\boldsymbol{r}, t)$ as $\phi_0(\boldsymbol{k}, t) = (1/V) \int d\boldsymbol{r} \exp(-i\boldsymbol{k} \cdot \boldsymbol{r})\phi_0(\boldsymbol{r}, t)$, we know that $\lim_{\boldsymbol{k} \to 0} \phi_0(\boldsymbol{k}, t)$ is in fact independent of $t$. Consequently, the temporal Fourier transform of $\phi_0(\boldsymbol{k}, t)$, defined by $\phi_0(\boldsymbol{k}, \omega) = (1/2\pi) \int dt \exp(-i\omega t)\phi_0(\boldsymbol{k}, t)$, has a property

$$\lim_{\boldsymbol{k} \to 0} \phi_0(\boldsymbol{k}, \omega) \sim \delta(\omega), \tag{13.17}$$

which implies that in the long wavelength limit there must be excitations whose frequency is zero in the Bose condensate. These massless excitations are nothing but Goldstone modes. As is well known, the Goldstone bosons in superfluid He II are identified as second sound. By means of Goldstone's theorem one can conclude that in the Bose condensate the gauge symmetry is broken, showing that the broken gauge symmetry and phase coherence are intimately connected.

It can be seen from the preceding discussions that superconductivity and superfluidity are the macroscopic phenomena of the Fermi and Bose systems, respectively, characterized by possessing the ODLRO. Both have many corresponding similarities, as summarized in the following table.

Table 13.1

| Superfluid | Superconductor |
|---|---|
| Coriolis force | Electromagnetic force |
| Angular velocity $\boldsymbol{\Omega}$ | Magnetic field $\boldsymbol{B}$ |
| Superflow velocity $\boldsymbol{v}_s(\boldsymbol{r})$ | Magnetic vector potential $\boldsymbol{A}(\boldsymbol{r})$ |
| ODLRO in $\rho_1$ | ODLRO in $\rho_2$ |
| Irrotational flow | Meissner effect |
| $\boldsymbol{\Omega} = \dfrac{1}{2}\nabla \times \boldsymbol{v}_s = 0$ | $\boldsymbol{B} = \nabla \times \boldsymbol{A} = 0$ |
| Quantization of vortices | Quantization of magnetic flux |
| $\oint \boldsymbol{v}_s \cdot d\boldsymbol{l} = n\dfrac{h}{m}$ | $\oint \boldsymbol{A} \cdot d\boldsymbol{l} = n\dfrac{ch}{2e}$ |
| Broken U(1)gauge symmetry | Broken U(1) gauge symmetry |
| Phase coherence | Phase coherence |
| Bose-Einstein condensate | Condensate of Cooper pairs |
| Macroscopic quantum state | Macroscopic quantum state (BCS) |
| $\Phi_0(\boldsymbol{r}) = \phi_0(\boldsymbol{r})e^{i\theta(\boldsymbol{r})}$ | $|\Psi_{\text{BCS}}\rangle = \Pi_k(u_k + v_k c_{k\uparrow}^\dagger c_{-k\downarrow}^\dagger)|\text{vac}\rangle$ |

## 13.4    ODLRO in Quantum Systems

The ODLRO exists in some quantum systems. For the well-known BCS superconducting state $|\Psi_{\text{BCS}}\rangle$ [7], it can be proved that it possesses the ODLRO of the two-particle reduced density matrix. As $|r - r'| \to \infty$, we have

$$\langle r, r|\rho_2|r', r'\rangle = \langle \Psi_{\text{BCS}}|c_{r\uparrow}^{\dagger}c_{r\downarrow}^{\dagger}c_{r'\downarrow}c_{r'\uparrow}|\Psi_{\text{BCS}}\rangle \to \langle c_{r\uparrow}^{\dagger}c_{r\downarrow}^{\dagger}\rangle\langle c_{r'\downarrow}c_{r'\uparrow}\rangle \neq 0, \tag{13.18}$$

where $|\Psi_{\text{BCS}}\rangle = \Pi_k(u_k + v_k c_{k\uparrow}^{\dagger}c_{-k\downarrow}^{\dagger})|\text{vac}\rangle$, $c_{k\sigma}^{\dagger}$ is the creation operator of an electron with momentum $k$ and spin $\sigma$, $u_k$ and $v_k$ are the functions introduced in BCS theory, $|\text{vac}\rangle$ is the vacuum state of electrons. This result shows that the BCS wave function has a property of cluster separation in the ODLR limit, and possesses the ODLRO.

In 1989, C. N. Yang proposed an $\eta$ pairing mechanism for the Hubbard model [40], and found that an eigenstate of the Hubbard model, the so-called $\eta$ pairing state, has the form of

$$|\Psi_N\rangle = \beta(\eta^{\dagger})^N|\text{vac}\rangle, \tag{13.19}$$

where the $\eta$ pairing operator is defined by $\eta^{\dagger} = \sum_k c_{k\uparrow}^{\dagger}c_{\pi-k\downarrow}^{\dagger} = \sum_r (-1)^r c_{r\uparrow}^{\dagger}c_{r\downarrow}^{\dagger}$, $\beta$ is the normalization factor, and $2N$ is the total number of electrons. It was proven that $|\Psi_N\rangle$ has the ODLRO of $\rho_2$ [40],

$$\langle \rho_2\rangle = \langle \Psi_N|c_{r\uparrow}^{\dagger}c_{r\downarrow}^{\dagger}c_{r'\downarrow}c_{r'\uparrow}|\Psi_N\rangle = \frac{N(M - N)}{M(M - 1)}e^{i\pi\cdot(r-r')} \quad (r \neq r'), \tag{13.20}$$

where $M$ is the total number of lattice sites. As $|r-r'| \to \infty$, $\langle \rho_2\rangle \neq 0$, showing that $|\Psi_N\rangle$ has the ODLRO of $\rho_2$. It should be noted that the $\eta$ pairing state $|\Psi_N\rangle$ is not the ground state of the Hubbard model, and is metastable. Very recently, there has been a proposal to measure the $\eta$ pairing state by means of the cold atoms on the optical lattice [41].

However, it has been proven that neither the two-dimensional (2D) Hubbard model [42, 43] nor the 2D $t$-$J$ model [44] possesses the $s$- and $d$-wave pairing ODLRO at finite temperature.

The discovery of the pseudogap phenomenon in underdoped cuprate oxide superconductors renewed the interest of possible BEC of electron pairs. One point of view believes that the BEC of preformed electron pairs may be related to the pseudogap phenomenon, because in such SC materials the coherence length is very small, and the system may be smoothly crossover from the weakly coupled BCS regime to the strongly coupled BEC regime, leading to a BCS-BEC crossover. On the other hand, in experiments of cold atoms, with the help of Feshbach resonance the BCS-BEC crossover can be simulated. In 1980, Leggett used a variational method and found that [45], at zero temperature the weakly coupled BCS superconducting ground state can evolve smoothly into the BEC condensate of tightly bound electron pairs (similar to a two-atom molecule). This scheme was extended and explored by many authors. In the strong coupling limit, the tightly bound electron pairs were usually treated as bosons. However, a closer check showed that such an equivalence is somewhat strained, because the tightly bound electron pairs are not exactly bosons. Our study [20] showed that the BEC of the tightly bound electron pairs may not directly occur in momentum space. If two bosonic quasiparticles, coined as binon and vacanon, are introduced, the BEC of the tightly bound electron pairs can be realized through these two kind of quasiparticles. A binon denotes the fluctuations of the tightly bound electron pairs, and a vacanon corresponds to a vacant state. By projecting the whole Hilbert space onto a subspace spanned by the states of binons and vacanons without doubly occupied states, it can be found that the hopping process of the tightly bound electron pairs is associated with the opposite hopping processes of binons and vacanons. By applying this idea to the strongly attractive Hubbard model, within the linearized approximation, we found that the attractive Hubbard model can be mapped onto a noninteracting system of binons and vacanons in the strong coupling limit [20]. For a state of $N_a$ binons, given by

$$|\phi(\boldsymbol{k})\rangle = \frac{1}{\sqrt{N_a!}}[a^\dagger(\boldsymbol{k})]^{N_a}|0\rangle, \qquad (13.21)$$

where $a^\dagger(\boldsymbol{k})$ is the Fourier transform of a binon field $\psi_a^\dagger(\boldsymbol{r})$, one may verify that $|\phi(\boldsymbol{k})\rangle$ is an eigenstate of the linearized Hamiltonian. In addition, it can be shown that

$$\langle\phi(\boldsymbol{k})|\psi_a^\dagger(\boldsymbol{r})\psi_a(\boldsymbol{r}')|\phi(\boldsymbol{k})\rangle = \frac{N_a}{M}e^{-i\boldsymbol{k}\cdot(\boldsymbol{r}-\boldsymbol{r}')}. \qquad (13.22)$$

If $N_a$ is an extensive quantity on the order of $O(M)$, then $\lim_{|r-r'|\to\infty}\langle\phi(\boldsymbol{k})| \psi_a^\dagger(\boldsymbol{r})\psi_a(\boldsymbol{r}')|\phi(\boldsymbol{k})\rangle = O(1) \neq 0$, implying that the binon state $|\phi(\boldsymbol{k})\rangle$ has the

ODLRO of the single-particle reduced density matrix, thereby exhibiting the BEC. Consequently, it appears that we can understand qualitatively the BCS-BEC crossover with the coupling constant in the attractive Hubbard model in such a picture. In the weak coupling limit, the system enters into the BCS regime, characterized by the formation of Cooper pairs; when the coupling is smoothly increased to intermediate values, the system enters into the crossover regime, where the Cooper pairs begin to be broken up, and the charge fluctuations become dominant; when the coupling is strong, the tightly bound electron pairs begin to form, and the system enters into the BEC regime, where the binons and vacanons play a crucial role in the formation of BEC. The concepts of binon and vacanon can also be applied to the Fermi cold atoms, boson-fermion mixtures, and Schafroth's SC theory of the tightly bound electron pairs [46], and so on.

When a system possesses simultaneously both the ODLRO and the diagonal long range order (DLRO) such as the translational invariance in a periodic crystal, charge density wave (CDW), and so forth, it is dubbed as a supersolid, which has been proposed and studied for many years [47]. According to the definition, a superconductor is a trivial supersolid. In 2004, Kim and Chan observed an unusual decoupling of the solid helium from a container's walls, i.e., the rotational inertia exhibits a non-classical behavior, and believed that it might indicate the solid $^4$He is a supersolid [48]. The subsequent theoretical and experimental works showed that Kim and Chan's experimental observation was not sufficient to make sure that the solid $^4$He is a supersolid, which is still in debate.

Quite recently, we have investigated a three-dimensional (3D) extended Bose Hubbard model, and obtained accurately the global phase diagram of the ground state [49]. The Hamiltonian of this model reads

$$\hat{H} = -t \sum_{\langle i,j \rangle} (\hat{b}_i^\dagger \hat{b}_j + \hat{b}_j^\dagger \hat{b}_i) + \frac{U}{2} \sum_i \hat{n}_i(\hat{n}_i - 1)$$
$$+ V \sum_{\langle i,j \rangle} \hat{n}_i \hat{n}_j - \mu \sum_i \hat{n}_i, \tag{13.23}$$

where $\hat{b}_i^\dagger$ ($\hat{b}_i$) is the creation (annihilation) bosonic operator at site $i$, $t$ is the hopping amplitude, $\hat{n}_i = \hat{b}_i^\dagger \hat{b}_i$ is the particle number, $\mu$ is the chemical potential, and $\langle i,j \rangle$ runs over all nearest neighbors. It is found that there are five phases, including superfluid, supersolid, CDW I, CDW II, and Mott insulating phases in its ground state, as shown in Fig. 13.1. By changing the chemical potential, we have revealed that the phase transition from the Mott phase to the superfluid phase is continuous without breaking the translational symmetry; an insulating CDW state by adding particles can be changed through a continuous transition into a supersolid phase, while a first-order transition to either a supersolid or superfluid phase could happen by removing particles. By tuning the nearest-neighbor interaction, the first-order phase transition between the Mott and CDW insulating phases with the same particle density can occur. A few years ago, a theoretical suggestion to realize a supersolid in a one-dimensional (1D) crystal by means of the cold atoms in an optical lattice was proposed [50]. We believe that, by utilizing the cold atoms in a

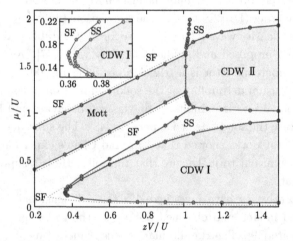

Fig. 13.1 The global phase diagram in the ground state of the 3D extended Bose Hubbard model [49], where $\mu$ is the chemical potential, $V$ is the next nearest-neighbor interaction, $z$ is the coordinate number, the on-site Coulomb interaction $U = 40t$, $t$ is the hopping integral, and the the size of the cubic lattice is taken as $12 \times 12 \times 12$. SF: superfluid phase; SS: supersolid phase; Mott: Mott insulating phase; CDW I and CDW II: charge density wave phase I and II. Solid lines denote the results obtained by the quantum Monte Carlo calculations, while the dotted lines represent the results from the perturbative calculations. Inset is the enlarged part of the lobe in the CDW I phase.

3D optical lattice, and by tuning the next nearest neighbor interaction and adjusting the other parameters to proper regimes, the supersolid described by the 3D extended Bose Hubbard model can be realized in experiments, which really deserves the experimentalists in the field of cold atoms to pay attention.

A different kind of ODLRO was shown to exist in the ground state of the fractional quantum Hall effect [51]. A further study manifested that such an ODLRO appears in the incompressible states of the quantum Hall effect with the filling fraction $\nu = 1/3$ and $\nu = 2/5$, while in compressible states there is no such an ODLRO [52].

Another kind of ODLRO was suggested in quantum spin chains [53], which is associated with the operators that add or delete sites from the spin chains. For the 1D periodic Heisenberg and the inverse square exchange Heisenberg models (both belong to the same universality class), it was shown that the overlap between the ground states of these two models with $N$ sites and the ground states with $(N - 2)$ sites plus a pair of nearest neighbor spins with a singlet configuration is nonzero, characterized by a hidden ODLRO, suggesting that long range correlations exist between the singlet-insertion and the singlet-deletion states, which destroy the $Z_2$ symmetry.

Besides, it was proved that the ODLRO also exists in a 1D exactly solvable many-body quantum model [54]. Moreover, in elementary particle physics, the condensation of quarks, color superconductivity, etc., may be also closely related to the existence of ODLRO.

## 13.5   Concluding Remarks

We may see that superconductor, superfluid, supersolid, and incompressible fractional quantum Hall state, etc., exhibit the ODLRO. Therefore, it seems that we may introduce a new concept, say, *superstate*, to overall describe the condensed states that possess the ODLRO. From the preceding discussions, one may find that different superstates usually demonstrate the same or similar properties, and have corresponding physical quantities and phenomena. It deserves to mention that the nontrivial supersolid is one of the superstates that are not definitely discovered up to date, although some hints and evidences were reported in literature, and the experimentalists could pay more attention on such a superstate. There have been great many studies on superconductors, superfluids, fractional quantum Hall effect, and so on, but many basic and challenging issues still remain unsolved.

One hundred years have passed since the discovery of superconductivity. Seeking for room temperature superconductors is still the dream of condensed matter physicists and materials scientists. Recently, a few conditions to obtain high temperature superconductors were suggested [55]: the material should have strong screened Coulomb interactions, strong dissipations, avoiding the repulsions between electrons, and spatial inhomogeneities. Whether the materials that satisfy these conditions are superconducting or not, still needs to attempt and examine.

The SC gap in a traditional superconductor opens only below the transition temperature, but a pseudogap in an underdoped cuprate oxide superconductor already opens above the SC critical temperature. The issues like what is really the pseudogap state, what is the relationship between the pseudogap and SC states and, what is the electron pairing mechanism (glue) in cuprate superconductors, etc., are still the active topics of experimental as well as theoretical studies currently. At present there are two influential views. One point of view believes that the pseudogap is the state of preformed electron pairs. Those preformed electron pairs somehow appear just below the so-called crossover temperature $T^*$, and owing to the thermal fluctuations, the coherent condensation of the preformed electron pairs does not occur. When the temperature is cooled down below the SC transition temperature $T_c$, the thermal fluctuations are suppressed, and the preformed electron pairs begin to condense coherently, leading to superconductivity. This is the one-gap scenario. Another point of view assumes that, the pseudogap differs quite from the SC gap, which is formed due to a kind of competing order with the SC phase, like charge density wave, spin density wave, $d$-density wave, electron pair density wave, etc. Such a competing order emerges just below the temperature $T^*$, and continues to coexist with the SC state even below the critical temperature $T_c$. This is the two-gap scenario. Which scenario, either one-gap or two-gap, will eventually win the bout, still needs more experimental inputs. Three experimental means are primarily involved to measure the pseudogap, namely, angle-resolved photoemission spectroscopy (ARPES), scanning tunneling microscopy (STM), and electronic Raman spectroscopy (ERS). Because various measuring methods focus on different points and use disparate mapping ways, different or even contrary experimental results are usually obtained. However, most observations come to support such a fact that there should exist two energy scales in the hole-doped cuprate superconductors at least in underdoped regimes, which are related to the nodal and antinodal regions in momentum space. Whether there is a relation between these two energy scales and the existence of the competing order with the SC phase is still under debate, and more experimental efforts are needed. A recent observation shows that in the pseudogap state the competing order coexists with the preformed electron pairs [56]. The diversity of the experimental results also leads to rich models and proposals, such as $d$-density wave order, the Fermi surface nesting in antinodal regions, electron pair fluctuations, resonant valence bond, electron pair density wave or $d$-wave checkerboard order, and coexistence of multiple competing orders in the pseudogap state, etc. In addition, there still lacks the well-accepted experimental evidence to determine where the glue force that can result in the electron pairing in cuprate high temperature superconductors comes from. The relationship between the symmetries of the SC gap and the pseudogap and the topology of the Fermi surface is also elusive. In unconventional BCS superconductors, the topology of Fermi surface and the symmetry of the gap function are indeed related [57]. For the iron-based superconductors, it is usually thought that the magnetic fluctuations participate directly or indirectly in the formation of electron pairs, but there are also theories assuming that the pairing mechanism is a combined result of the Coulomb repulsion between electrons, spin fluctuations and electron-phonon couplings. The consensus is not yet reached on the key issues such as the SC symmetry, itinerant or localized magnetic moments, the role of orbital degree of freedom, strongly or weakly correlated electrons, the concise form of the Hamiltonian, etc., in these Fe-based materials, which still need a great deal of inputs. There is a recent study [58] showing that in the iron chalcogenides, there exists a re-emerging, second SC phase above the pressure 11.5 GPa. Therefore, to overall well explain the high temperature superconductivity in Cu- and Fe-based materials, there is a long way to go, and more experimental and theoretical efforts should be made. Nevertheless, no matter what a superconducting theory for high temperature superconductivity is eventually accepted, the ODLRO would play an indispensable role.

Last but not least, I would like to add the open questions and future directions for exploring the superconductivity, which I believe are very important and interesting, for instance, among others, the microscopic mechanism of high temperature superconductivity (including the pairing mechanism, the form of Hamiltonian, the origin of pseudogap, the relationship between the topology of Fermi surface and the pairing symmetry, etc.); searching for new superconductors; seeking for room temperature superconductors; developing new theories beyond the BCS; finding novel applications of superconductors; performing the interdisciplinary research; etc. We have reasonable ground to imagine that, after another one hundred years of explorations on superconductivity, the living styles of human beings would be much more benefitted from superconductors than ever.

## Acknowledgments

I am indebted to Professor Mo-Lin Ge, Professor Hwa-Tung Nieh, Professor Chen-Ning Yang, and Professor Bao-Heng Zhao for enlightening suggestions and discussions over the past years. This work is supported in part by the NSFC and the Chinese Academy of Sciences.

[1]  H. K. Onnes, Commun. Phys. Lab. Univ. Leiden **12**, 120 (1911).

[2]  W. Meissner and R. Ochsenfeld, Naturwissenschaften **21**, 787 (1933).

[3]  F. London and H. London, Proc. Roy. Soc. London A **149**, 71 (1935).

[4]  V. L. Ginzburg and L. D. Landau, Zh. Eksp. i Teor. Fiz. **20**, 1064 (1950).

[5]  A. A. Abrikosov, Zh. Eksp. i Teor. Fiz. **32**, 1442 (1957).

[6]  E. Maxwell, Phys. Rev. **78**, 477 (1950); C. A. Reynolds, B. Serin, W. H. Wright and L. B. Nesbitt, Phys. Rev. **78**, 487 (1950).

[7]  J. Bardeen, L. N. Cooper and J. R. Schrieffer, Phys. Rev. **108**, 1175 (1957).

[8]  G. M. Eliashberg, Zh. Eksp. i Teor. Fiz. **38**, 966 (1960).

[9]  B. D. Josephson, Phys. Lett. **1**, 251 (1962).

[10]  J. G. Bednorz and K. A. Mueller, Z. Phys. B **64**, 189 (1986).

[11]  Y. Kamihara, T. Watanabe, M. Hirano and H. Hosono, J. Am. Chem. Soc. **130**, 3296 (2008).

[12]  N. Imamura, H. Mizoguchi and H. Hosono, J. Am. Chem. Soc. **134**, 2516 (2012).

[13]  N. W. Ashcroft, Phys. Rev. Lett. **21**, 1748 (1968).

[14] P. Tripodi, D. D. Gioacchino, R. Borelli and J. D. Vinko, Physica C **388-389**, 571 (2003); Int. J. Mod. Phys. B **21**, 3343 (2007).

[15] http://www.superconductors.org/20C.htm.

[16] Y. Zheng and G. Su, Science in China G: Phys. Mech. Astro. **39** (11), 1553-1570 (2009).

[17] G. Su, in *Experiments and Theories of Cuprate High Temperature Superconductivity*, edited by R. S. Han, H. H. Wen and T. Xiang (Science Press, Beijing, 2009), pp. 259-273.

[18] S. N. Bose, Z. Phys. **26**, 178 (1924).

[19] A. Einstein, Sitzungsberichte der Preussischen Akademie der Wissenschaften **1**, 3 (1925).

[20] G. Su and M. Suzuki, Int. J. Mod. Phys. B **13**, 925 (1999).

[21] G. Su and M. Suzuki, Eur. Phys. J. B **5**, 577 (1998).

[22] F. London, Nature **141**, 643 (1938).

[23] L. Tisza, Nature **141**, 913 (1938); Phys. Rev. **72**, 838 (1947).

[24] L. D. Landau, J. Phys. USSR **5**, 71 (1941).

[25] L. Onsager, Nuovo Cimento **6**, Suppl. 2, 249 (1949).

[26] R. P. Feynman, in *Progress in Low Temperature Physics*, Vol. 1, edited by C. J. Gorter (North-Holland, Amsterdam, 1955).

[27] H. E. Hall and W. F. Vinen, Proc. Roy. Soc. A **238**, 204; 215 (1956).

[28] M. H. Anderson, J. R. Ensher, M. R. Matthews, C. E. Wieman and E. A. Cornell, Science **269**, 198 (1995).

[29] K. B. Davis, M. O. Mewes, M. R. Andrews, N. J. Van Druten, D. S. Durfee, D. M. Kurn and W. Ketterle, Phys. Rev. Lett. **75**, 3969 (1995).

[30] B. T. Seaman, M. Krämer, D. Z. Anderson and M. J. Holland, Physical Review A **75**, 023615 (2007); R. A. Pepino, J. Cooper, D. Z. Anderson and M. J. Holland, Phys. Rev. Lett. **103**, 140405 (2009).

[31] N. N. Bogoliubov, J. Phys. U.S.S.R. **11**, 23 (1947).

[32] O. Penrose, Philos. Mag. **42**, 1373 (1951); O. Penrose and L. Onsager, Phys. Rev. **104**, 576 (1956).

[33] C. N. Yang, Rev. Mod. Phys. **34**, 694 (1962).

[34] J. Goldstone, Nuovo Cimento **19**, 154 (1961).

[35] P. W. Anderson, Rev. Mod. Phys. **38**, 298 (1966).

[36] T. L. Ho and V. B. Shenoy, Phys. Rev. Lett. **77**, 2595 (1996); T. L. Ho, Phys. Rev. Lett. **81**, 742 (1998).

[37] G. L. Sewell, J. Stat. Phys. **61**, 415 (1990); J. Math. Phys. **38**, 2053 (1997).

[38] H. T. Nieh, G. Su and B. H. Zhao, Phys. Rev. B **51**, 3760 (1995).

[39] G. Su and M. Suzuki, Phys. Rev. Lett. **86**, 2708 (2001).

[40] C. N. Yang, Phys. Rev. Lett. **63**, 2144 (1989).

[41] A. Kantian, A. J. Daley and P. Zoller, Phys. Rev. Lett. **104**, 240406 (2010).

[42]  G. Su and M. Suzuki, Phys. Rev. B **58**, 117 (1998).

[43]  G. Su, Phys. Rev. Lett. **86**, 3690 (2001).

[44]  G. Su, Phys. Rev. B **72**, 092510 (2005).

[45]  A. J. Leggett, in *Modern Trends in the Theory of Condensed Matter*, edited by A. Pekalski and R. Przystawa (Springer-Verlag, Berlin, 1980).

[46]  M. R. Schafroth, Phys. Rev. **100**, 463 (1955).

[47]  A. F. Andreev and I. M. Lifshitz, Sov. Phys. JETP **29**, 1107 (1969); A. J. Leggett, Phys. Rev. Lett. **25**, 1543 (1970); G. V. Chester, Phys. Rev. A **2**, 256 (1970).

[48]  E. Kim and M. H. Chan, Science **305**, 1941 (2004); Nature **427**, 225 (2004).

[49]  B. Xi, F. Ye, W. Q. Chen, F. C. Zhang and G. Su, Phys. Rev. B **84**, 054512 (2011).

[50]  T. Keilmann, I. Cirac and T. Roscilde, Phys. Rev. Lett. **102**, 255304 (2009).

[51]  S. M. Girvin and A. H. MacDonald, Phys. Rev. Lett. **58**, 1252 (1987).

[52]  E. H. Rezayi and F. D. M. Haldane, Phys. Rev. Lett. **61**, 1985 (1988).

[53]  J. C. Talstra, S. P. Strong and P. W. Anderson, Phys. Rev. Lett. **74**, 5256 (1995).

[54]  G. Auberson, S. R. Jain and A. Khare, Phys. Lett. A **267**, 293 (2000).

[55]  D. N. Basov and A. V. Chubukov, Nat. Phys. **7**, 272 (2011).

[56]  T. Kondo, Y. Hamaya, A. D. Palczewski, T. Takeuchi, J. S. Wen, Z. J. Xu, G. D. Gu, J. Schmalian and A. Kaminski, Nat. Phys. **7**, 21 (2011).

[57]  G. Su and M. Suzuki, Int. J. Mod. Phys. B **14**, 563 (2000).

[58]  L. Sun *et al.*, Nature **483**, 67 (2012).

# 14

# Superconductivity of Cuprates—A Phenomenon of Strong Correlation of Electrons*

W. Guo and R. S. Han

*School of Physics, Peking University, Beijing 100871, China*

The strong correlation of electrons such as the magnetic ordering in transition-metal-oxides and the Kondo resonance in dilute magnetic compounds rising from the local exchange interactions were well understood many-body effects in the past decades. The superconducting cuprate discovered in 1986 inspired a fascinating idea: the magnetic origin of superfluid. We speculate that the fluctuating spins on the oxygen sites in the $CuO_2$ layer created by the holes may pair into a resonating-valence-bond state with the quantum number $S = 1, S_z = 0$ via local exchange interactions, the same force responsible for magnetic ordering in the low doped cuprate. Impurity doping drives the cuprate from an antiferromagnetic insulator to a superconductor, then a Fermi liquid giving a rich phase diagram.

The superconducting cuprates discovered by Müller and Bednorz in 1986 are essentially different from the conventional BCS superconductors, they are strongly correlated systems characterized by the coexistence of superconductivity and magnetism [1–8]. Like many of the transition-metal-oxides the undoped cuprate is an antiferromagnetic insulator, when the cuprate is doped with impurity it becomes a superconductor with a coherent length about 10 Å. The short coherent length is an indication of the local pairing. Is the same interaction responsible for both of the magnetic ordering and the electron

---

* Based on the talk presented at the Symposium on High $T_c$ Superconductivity, 2008 CCAST.

pairing? A classic example for strong correlation between itinerant charge carriers and local spins is the Kondo resonance where the conduction electron spins and a local spin form a resonating singlet ground state via $s$-$d$ exchange interaction. In case of the high $T_c$ problem, we focus on the issue of the formation of the resonating-valence-bond state, a key concept to understanding superconductivity in cuprates.

## 14.1   Electronic Structure of the Cuprate and Magnetism

The structure unit responsible for superconductivity in a cuprate is the $CuO_2$ layer, in which the copper atoms at each corner of a single square are separated by an oxygen atom (Fig. 14.1). The Cu site is singly occupied by a $d$ electron with $S = 1/2$ since the large on-site Coulomb interaction $U$ splits the half filled $d$ band into two sub-bands, i.e. the Hubbard bands. The O site is nonmagnetic where the $p$ orbital is doubly occupied, the Hartree-Fork approximation applies, the filled $p$ band sits between two Hubbard bands of $d$ electrons [9]. The $CuO_2$ layer is insulating for an undoped cuprate.

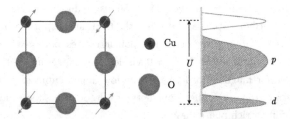

Fig. 14.1   The magnetic and the electronic structures of the $CuO_2$ layer.

The basic model describing the interactions and the electronic states of the $CuO_2$ layer in a cuprate is the two band model:

$$H = \sum_{l\sigma} \varepsilon_p p_{l\sigma}^\dagger p_{l\sigma} + \sum_{i\sigma} \varepsilon_d d_{i\sigma}^\dagger d_{i\sigma} + U d_{i\uparrow}^\dagger d_{i\uparrow} d_{i\downarrow}^\dagger d_{i\downarrow}$$
$$+ \sum_{i,l\in\{i\},\sigma} V_{il} \left( d_{i\sigma} p_{l\sigma}^\dagger + d_{i\sigma}^\dagger p_{l\sigma} \right), \qquad (14.1)$$

where $d_{i\sigma}^\dagger, d_{i\sigma}$, $p_{l\sigma}^\dagger$, $p_{l\sigma}$ are creation and annihilation operators for $d$ electrons at Cu sites $i$, and $p$ electrons at O sites $l$, $l \in \{i\}$ are the nearest neighbors of the Cu site $i$. $U$ is the on-site Coulomb interaction for $d$ electrons. For undoped cuprates, one can derive the effective spin interaction of Cu-site spins by using perturbation theory, known as Kramers superexchange theory where the $p$ electrons play the role of mediation for the magnetic interaction between Cu-site spins [10] (Fig. 14.2),

$$H = -K \sum_{i,l \in \{i\}} S_i \cdot S_j, \qquad (14.2)$$

$K < 0$ is the antiferromagnetic coupling constant,

$$K \propto \rho^2 J_d,$$

where $\rho$ is the transition integral $p \to d$ transition,

$$\rho = \int d^3 r \phi_d(r) \hat{V} \phi_p(r).$$

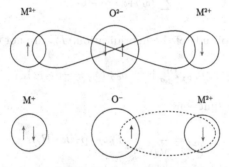

Fig. 14.2  Kramers superexchange. The ground state (above) and the intermediate states (below) relevant in the perturbation calculations. For the excited states, an O spin jumps to the Cu site on one side, the other O spin interacts with the Cu spin on the other side via direct exchange.

The $d$ orbital $\phi_d$ in a $Cu^{2+}$ ion is a localized state with the $d_{x^2-y^2}$ symmetry, the $p$ state on the other hand is described by a Wannier wavefunction $\phi_p$ in 2D dimension. The coupling between $p$ and $d$ electrons arises from the overlap of $\phi_d$ and $\phi_p$, the magnetization of Cu spins in the $CuO_2$ layer is observed in neutron scattering experiments [11].

## 14.2    The Magnetic Origin of Superconductivity in Cuprates

Anderson proposed in 1987 that the spin $S = 1/2$ lattice described by Eq. 14.2 has a type of the ground state where the local spins have no preferential directions, the neighboring Cu spins are paired into the singlets, so-called the resonating-valence-bond state, and condense to the superconducting state when the doped $CuO_2$ layer becomes metallic [12].

In fact, the doped cuprate has a two-band structure. The holes introduced in the $CuO_2$ layer reside on O sites. In this case, whether the degrees of freedom of the $p$ electrons are integrable remains a question. In order to derive an effective single band model for the superconducting cuprates, Zhang and Rice proposed that the hybridization between $p$ and $d$ electrons binds a hole on each square of O atoms to the central $Cu^{2+}$ ion to form a local singlet [13]. Zhang-Rice started from the two-band model (Eq. 14.1) where $p_i^\dagger$ and $p_i$ are creation and annihilation operators for holes. The four oxygen hole states around a Cu ion were written in a symmetric form,

$$\sum_\delta t_\delta p_{i+\delta,\sigma} = 2t_0 P_{i\sigma},$$

where $t_0$ is the amplitude of the hybridization. $P_{i\sigma}$ are not orthogonal,

$$[P_{i,\sigma}, P_{j,\sigma'}^\dagger]_+ = \delta_{\sigma\sigma'}(\delta_{ij} - \frac{1}{4}\delta_{<ij>}).$$

We define Wannier functions,

$$\phi_{i\sigma} = \frac{1}{\sqrt{N}} \sum_k P_{k\sigma} \exp{(i\boldsymbol{k} \cdot \boldsymbol{R}_i)},$$

$$P_{k\sigma} = \frac{1}{\sqrt{N}} \beta_k \sum_i P_{i\sigma} \exp{(-i\boldsymbol{k} \cdot \boldsymbol{R}_i)},$$

where $\beta_k$ is a normalization factor. $\phi_{i\sigma}$ is orthogonal,

$$[\phi_{i\sigma}, \phi_{j\sigma'}^\dagger]_+ = \delta_{\sigma\sigma'}\delta_{ij}.$$

$P_{i\sigma}$ can be expanded as

$$P_{i\sigma} = \sum_j \lambda(R_i - R_j)\phi_{j\sigma} = \lambda_0 \phi_{i\sigma} + \lambda_1 \sum_{<1>} \phi_{<1>\sigma} + \lambda_2 \sum_{<2>} \phi_{<2>\sigma} + \cdots, \quad (14.3)$$

where $\lambda_0 \approx 0.96$, $< 1 >$ are the nearest neighbor sites around site $i$ with $\lambda_1 \approx 0.14$, $< 2 >$ are the next nearest, we have

$$H = \sum_{i,\sigma} \varepsilon_d d_{i\sigma}^\dagger d_{i\sigma} + U \sum_i n_{i\uparrow} n_{i\downarrow} + \lambda_0^2 \sum_{i,\sigma} \phi_{i\sigma}^+ \phi_{i\sigma} + H', \qquad (14.4)$$

where $H'$ is the hybridization interaction

$$H' = 2\lambda_0 t_0 \sum_{i,\sigma} (d_{i\sigma}^+ \phi_{i\sigma} + h.c.) + 2\lambda_n t_0 \sum_{i,\sigma,n=1}^{\infty} (d_{i\sigma}^+ \phi_{<n>\sigma} + h.c.).$$

$H'$ is taken as perturbation term, the hybridization binds an O hole to a Cu at site $i$. Two states are relevant,

$$|\alpha\rangle = d_{i\uparrow}^\dagger \phi_{i\downarrow}^\dagger |B\rangle, \qquad |\beta\rangle = d_{i\downarrow}^\dagger \phi_{i\uparrow}^\dagger |B\rangle, \qquad (14.5)$$

where

$$|B\rangle = \sum_{l=1}^{M} a_l |B_l\rangle, \qquad |B_l\rangle = \prod_{j\neq i} d_{j\sigma_l(j)}^\dagger |0\rangle.$$

$|B_l\rangle$ is the spin configuration of the whole Cu spin lattice excluding site $i$. By using second order perturbation theory to calculate the binding energy,

$$E^\pm = -8t(1 \mp \lambda_0^2) < 0, \qquad t = \frac{t_0^2}{\lambda_0^2 \varepsilon_p - \varepsilon_d},$$

where $E^+$ and $E^-$ are the binding energies of spin triplet and singlet. The singlet is the ground state. The singlet effectively screens a Cu spin, and moves through the lattice of $Cu^{2+}$ ions as a "hole" (Fig. 14.3).

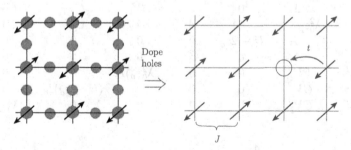

Fig. 14.3  Zhang-Rice's theory is an attempt to integrate the degrees of freedom for $p$ electrons reducing the two-band model to an effective single band model.

By integrating the degrees of freedom of $p$ electrons Zhang and Rice obtained an effective single band Hamiltonian

$$H_{t\text{-}J} = -t \sum_{ij\sigma} (c_{i\sigma}^\dagger c_{j\sigma} + h.c.) - J \sum_{ij} \boldsymbol{S}_i \cdot \boldsymbol{S}_j \tag{14.6}$$

to describe the low excitation states of the superconducting cuprate. Despite the simple appearance and the partial success in calculating for the low excitation states for the cuprates, the single band model failed to explain either the cross-over from the antiferromagnetic to the superconducting states as shown in the phase diagram, or coexistence of superconductivity and magnetism in the superconducting region.

## 14.3   Effect of the Local Spin Polarization

In the low doping limit the correlation of the local Cu spins in a cuprate is not negligible. We consider the competition between the local Cu spin polarization and Zhang-Rice's singlet to show that the Zhang-Rice singlet is decoupled by the arbitrarily weak spin polarization [14]. By introducing the local spin polarization

$$\Delta = \varepsilon_{d\uparrow} - \varepsilon_{d\downarrow}$$

to the two-band model (14.1), we adopt the diagonalization method. The number of states involving in second perturbation calculation is 28. We have a $28 \times 28$ matrix for $H$,

$$H - E_B =$$

$$\begin{pmatrix}
\lambda_0^2 \varepsilon_p + \varepsilon_d & \Delta & 0 & 0 & BV_1 & CV_1 & CV_1 \\
\Delta & \lambda_0^2 \varepsilon_p + \varepsilon_d & -A & A & BV_2 & CV_2 & CV_2 \\
0 & -A & U + 2\varepsilon_d & 0 & 0 & 0 & 0 \\
0 & A & 0 & 2\lambda_0^2 \varepsilon_p & 0 & 0 & 0 \\
BV_1^T & BV_2^T & 0 & 0 & (2\lambda_0^2 \varepsilon_p) I_8 & 0 & 0 \\
CV_1^T & CV_2^T & 0 & 0 & 0 & (U + 2\varepsilon_d) I_8 & 0 \\
CV_1^T & CV_2^T & 0 & 0 & 0 & 0 & (2\lambda_0^2 \varepsilon_p) I_8
\end{pmatrix},$$

$$\tag{14.7}$$

where

$$A = \frac{4}{\sqrt{2}} \lambda_0 t_0, \quad B = \frac{2}{\sqrt{2}} \lambda_1 t_0, \quad C = \lambda_1 t_0,$$

$$D_1 = \lambda_0^2 \varepsilon_p + \varepsilon_d, \quad D_2 = U + 2\varepsilon_d, \quad D_3 = \lambda_0^2 \varepsilon_p,$$

$I_8$ is a $8 \times 8$ unit matrix, $V_1$ and $V_2$ are $8 \times 1$ vectors,

$$V_1 = (\;1\;\;1\;\;1\;\;1\;\;1\;\;1\;\;1\;\;1\;), \quad V_2 = (\;1\;\;1\;\;1\;\;1\;-1\;-1\;-1\;-1\;).$$

The eigenvector of $H$ is defined as

$$\psi = (a_1,\;\; a_2,\;\; \cdots\;\; a_{28}), \tag{14.8}$$

where $a_1$ is the coefficient for the triplet state, $a_2$ is for the singlet. $a_3$ to $a_{28}$ are for the base vectors listed in Eq. 14.6, from which we have

$$(\lambda_0^2 \varepsilon_p + \varepsilon_d)a_1 + \Delta a_2 + B \sum_{n=5}^{12} a_n + C \sum_{n=13}^{28} a_n = E a_1,$$

$$B a_1 + B a_2 + 2\lambda_0^2 \varepsilon_p a_5 = E a_5,$$

$$B a_1 - B a_2 + 2\lambda_0^2 \varepsilon_p a_9 = E a_9.$$

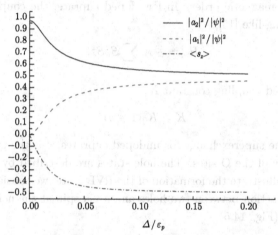

Fig. 14.4   The numerical computation for amplitudes of the triplet ($|a_1|^2$)and the singlet states ($|a_2|^2$) as functions of spin polarization $\Delta$, where we take $t/\varepsilon_p = 0.05, U/\varepsilon_p = 3$. The average spin on Cu site $\langle s_z \rangle = a_1 a_2$. Coefficient $a_2 = -a_1 \approx -1/\sqrt{2}$ at large $\Delta$, the eigenstate is a spin polarized state.

If there exists a singlet solution, it requires that $a_1 = 0$. It follows

$$\sum_{n=5}^{12} a_n = 0, \quad \sum_{n=13}^{28} a_n = 0,$$

so

$$\Delta \cdot a_2 = 0. \tag{14.9}$$

If $\Delta$ is nonzero, $a_2$ the coefficient for the singlet state equals to zero, thus the singlet state is unstable.

## 14.4   The Resonating-Valence-Bond State in the Superconducting Cuprates

In this section we show that the resonating-valence-bond (RVB) state can be generated in the two-band structure where holes are the charge carriers and Cu spins are localized providing a magnetic background. The holes introduced in $CuO_2$ layer not only induce metallic-insulating transition, but also suppress the on-site superexchange, the quasi-free spin flips on Cu sites breaking the long range antiferromagnetic order. In low doped cuprate, the coupling between Cu spins is Ising-like [15],

$$H_m = -K \sum_{ij} S_i^z S_j^z,$$

with a weakened coupling constant $K$,

$$K = K_0(1 - v),$$

where $K_0$ is the superexchange for undoped cuprates, $v$ is the average of the hole occupancy of the O sites. The hole states are described by the Wannier functions. To illustrate the formation of the RVB state, we consider the direct exchange interaction between two holes and a Cu spin at the neighbor site in a spin cluster (Fig. 14.5),

$$H_{\text{int}} = -J\boldsymbol{S} \cdot (\boldsymbol{\sigma}_1 + \boldsymbol{\sigma}_2), \quad J < 0,$$

where $J$ is the coupling constant of the direct exchange. $\boldsymbol{S}$ is the Cu spin. The polarized $\boldsymbol{S}$ has two positions: $|0\rangle$ the ground state and $|1\rangle$ the excited state. $\sigma_i$ $(i = 1, 2)$ are O spins. By using second order perturbation theory, we derive an effective spin coupling between O spins [16],

$$H_{\text{eff}} = \frac{\langle 0 | \hat{H}_{\text{int}} | 1\rangle \langle 1 | \hat{H}_{\text{int}} | 0\rangle}{E_0 - E_1},$$

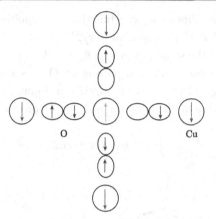

Fig. 14.5   The uncompensated spins created by holes residing on O sites in a spin cluster interact with the Cu spin at the center.

which is

$$H_{\text{eff}} = -\frac{1}{4}\lambda J(\sigma_1^+ + \sigma_2^+)(\sigma_1^- + \sigma_2^-), \tag{14.10}$$

where $\lambda = J/4K$, varies with the doped hole concentration, and $\sigma^{\pm} = \sigma_x \pm i\sigma_y$. $H_{\text{eff}}$ has an eigenstate

$$\chi_+ = \frac{1}{\sqrt{2}}\left(\mid \uparrow\downarrow\rangle + \mid \downarrow\uparrow\rangle\right),$$

with the binding energy

$$E_b = -\frac{1}{2}\lambda J\left(1 - \frac{1}{2}\lambda^2\right). \tag{14.11}$$

The spin pair has the quantum numbers of $S = 1, S_z = 0$, so that the possibility of $s$ wave pairing is excluded. For the local pairing, the symmetry of the electron pair is restricted by the symmetry of the crystal lattice of the $CuO_2$ layer, therefore the electron pair has been forced to have the $d$ symmetry [17]. We write an antisymmetric $d$ wavefunction in terms of scattered partial waves in $l = 2$ channel,

$$\psi = Y_2^{+1}(\theta_1, \varphi_1)Y_2^{-1}(\theta_2, \varphi_2) - Y_2^{-1}(\theta_1, \varphi_1)Y_2^{+1}(\theta_2, \varphi_2),$$

the total quantum number $m = m_1 + m_2 = 0$. Let $\theta = \theta_1 - \theta_2$, $\varphi = \varphi_1 - \varphi_2$,

$$\psi \propto \cos 2\theta \sin \varphi,$$

where $-\pi \leqslant \theta \leqslant \pi$. Since the wavefunction $\psi$ is limited in a 2D layer, we take a fixed value for $\varphi$, $\varphi \to \pi/2$, or $-\pi/2$(Fig. 14.6). The projection of $\psi$ in the $a$-$b$ plane has the same pattern as $d_{x^2-y^2}$, but it is consistent with the triplet pairing. The theory of spin pairing on O sites naturally explains the phase diagram (Fig. 14.7). Near the optimum doping, the superconducting transition temperature of the cuprates obey an empirical universal law,

$$\frac{T}{T_m} = 1 - \kappa(x - x_0)^2.$$

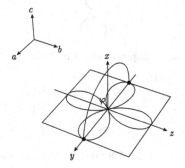

Fig. 14.6   The projection of the antisymmetric 3D wavefunction in the $a$-$b$ plane, where, $\varphi$ is defined in the $x$-$y$ plane.

From Eq. 14.11, we can derive the expression for $\kappa$,

$$\kappa = \left(\frac{4K_0}{J}\right)^2,$$

where $K_0$ is the superexchange constant of the undoped cuprate, $J$ is the direct exchange between the hole spins and a neighbor Cu spin. The upper limit of the superconducting transition temperature is determined by $K_0$,

$$T_{\max} = \frac{1}{6.81 k_B} \cdot \lambda_{\max} K_0.$$

If $K_0 \approx 0.1$ eV, $T_{\max} \sim 150$ K.

Based on the effective spin coupling Eq. 14.10, we propose an effective Hamiltonian for the superconducting cuprates,

$$H = \sum_{ij} t_{ij} p_i^\dagger p_j - g \sum_{i\delta\sigma} p_{i\sigma}^\dagger p_{i+\delta,-\sigma}^\dagger p_{i+\delta,\sigma} p_{i,-\sigma} - K \sum_{ij} S_i^z S_j^z, \qquad (14.12)$$

where $p_i^\dagger$ and $p_i$ are creation and annihilation operators of holes. The effective interaction Eq. 14.12 describes the local pair pairing in the presence of a magnetic background of Cu spins. The local pairs form a resonating-valence-bond state and condense, the cuprate becomes superconducting. $g$ depends on the doping level, when the $p$ band is filled, $g = 0$. Near the optimal doping, Eq. 14.12 gives a correct phase diagram.

$$g = \frac{J}{4K}.$$

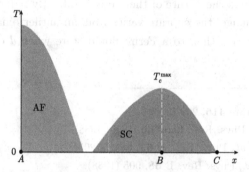

Fig. 14.7    The phase diagram of the cuprates. $A$, $B$ and $C$ are three critical doping levels: $A$, the undoped cuprate is an antiferromagnetic insulator where the $p$ electron plays a role of the mediation for the superexchange interaction for Cu spins. $B$, Near optimal doping, two O spins generated by the doped holes form an RVB pair via the effective spin coupling mediated by a Cu spin. $C$, the $d$ band electrons are delocalized and hybridized with the $p$ band electrons forming a Fermi liquid.

## 14.5    Summary

Based on Kramers' theory, we have shown the key role that the $p$ electrons play in superconductivity in a doped cuprate. In fact, Kramers' theory which describes the indirect interaction in the transition-metal-oxides where the $p$ electrons on O sites play the role of mediation is never tested in experiment by taking the $p$ electrons away until the superconducting cuprate is discovered. The phase diagram shows the antiferromagnetic order of Cu spins is sensitive to the impurity doping. The most important generation for Kramers' theory is the emergence of the resonating-valence-bond state when holes are introduced in $p$ orbital in a doped cuprate. The results of NMR experiments indicate the pairing occurs on O sites[18]. The $d$ electrons, which are emphasized in the single band model, merely provide magnetic background and play the role of mediation for $p$ electrons pairing. Two type of the interactions, the direct exchange $J$ between $p$ and $d$ electron spins and the superexchange $K$ between $d$ electron spins are essential to magnetic ordering and superconductivity in the cuprates. The upper limit of the superconducting transition temperature is determined by the strength of the superexchange of the undoped cuprate. Impurity doping alters the nature of the ground state. By reducing the strength of the superexchange, the cuprate varies from an antiferromagnetic insulator to a superconductor, then to a Fermi liquid state when $d$ electrons become delocalized.

[1]   B. Lake, Nature **415**, 299 (2002).

[2]   A. Aharony, Phys. Rev. Lett. **60**, 1330 (1988).

[3]   P. C. Dai *et al.*, Phys. Rev. B **63**, 54525 (2001).

[4]   T. Thio *et al.*, Phys. Rev. B **38**, 905 (1988).

[5]   M. Matsuda *et al.*, Phys. Rev. B **66**, 174508 (2002).

[6]   M. Takigawa, A. P. Reyes, P. C. Hammel, J. D. Thompson, R. H. Heffner, Z. Fisk and K. C. Ott, Phys. Rev. B **43**, 247 (1991).

[7]   A. Kaminski, S. Rosenkranz, H. M. Fretwell, J. C. Campuzano, Z. Li, H . Raffy, W. G. Cullen, H. You, C. G. Olson, C. M. Varma and H. Hochst, Nature **416**, 610 (2002).

[8]   W. Guo, L. S. Duan and R. S. Han, International Journal of Modern Physics B **19**, 63 (2005).

[9]   S. Uchida, T. Ido, H. Takagi, T. Arima, Y. Tokura and S. Tajima, Phys. Rev. B **43**, 7942 (1991).

[10]  Kramers, Physica **1**, 187 (1934).

[11]  D. V. Vaknin *et al.*, Phys. Rev. Lett. **58**, 280 (1987).

[12]  P. W. Anderson, Science **235**, 1196 (1987).

[13]  F. C. Zhang and T. M. Rice, Phys. Rev. B **37**, 3759 (1988).

[14]  H. Li and W. Guo, Journal of Superconductivity and Novel Magnetism **23**, 679 (2010).

[15]  A. N. Lavrov *et al.*, Phys. Rev. Lett. **87**, 017007 (2001).

[16]  W. Guo, X. G. Yin and R. S. Han, International Journal of Modern Physics B **21**, 3112 (2007).

[17]  C. C. Tsuei and J. R. Kirtley, Rev. Mod. Phys. **72**, 969 (2000).

[18]  G. Q. Zheng, Y. Kitaoka, K. Asayama, K. Hamada, H. Yamauchi and S. Tanaka, Physica C **260**, 197 (1996).

# 15

# Superconductivity at 43 K in SmFeAsO$_{1-x}$F$_x$*

X. H. Chen, T. Wu, G. Wu, R. H. Liu, H. Chen and D. F. Fang

*Hefei National Laboratory for Physical Sciences at Microscale and Department of Physics, University of Science and Technology of China, Hefei, Anhui 230026, China*

Since the discovery of high-transition-temperature (high-$T_c$) superconductivity in layered copper oxides, extensive effort has been devoted to exploring the origins of this phenomenon. A $T_c$ higher than 40 K (about the theoretical maximum predicted from Bardeen-Cooper-Schrieffer theory [1]), however, has been obtained only in the copper oxide superconductors. The highest reported value for non-copper-oxide bulk superconductivity is $T_c = 39$ K in MgB$_2$ [2]. The layered rare-earth metal oxypnictides LnOFeAs (where Ln is La-Nd, Sm and Gd) are now attracting attention following the discovery of superconductivity at 26 K in the iron-based LaO$_{1-x}$F$_x$FeAs [3]. Here we report the discovery of bulk superconductivity in the related compound SmFeAsO$_{1-x}$F$_x$, which has a ZrCuSiAs-type structure. Resistivity and magnetization measurements reveal a transition temperature as high as 43 K. This provides a new material base for studying the origin of high-temperature superconductivity.

LnO$_{1-x}$F$_x$FeAs adopts ZrCuSiAs-type structure. A series of equiatomic quaternary compounds LnFeAsO and LnFePO (Ln = La−Nd, Sm, Gd) with ZrCuSiAs-type structure has been reported [4,5]. The crystal structure of the tetragonal ZrCuSiAs-type compound SmFeAs(O,F) is shown in Fig. 15.1.

Polycrystalline samples with nominal composition SmFeAsO$_{1-x}$F$_x$ ($x =$ 0.15) were synthesized by conventional solid state reaction using high-purity

* Reprinted from X. H. Chen, T. Wu, G. Wu *et al.*, Nature **453**, 761 (2008).

SmAs, SmF$_3$, Fe and Fe$_2$O$_3$ as starting materials. SmAs was obtained by reacting Sm chips and As pieces at 600°C for 3 h and then 900°C for 5 h. The raw materials were thoroughly ground and pressed into pellets. The pellets were wrapped in Ta foil, sealed in an evacuated quartz tube, and finally annealed at either 1160°C or 1200°C for 40 h.

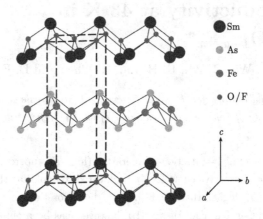

Fig. 15.1 Structural model of SmFeAsO$_{1-x}$F$_x$ with the tetragonal ZrCuSiAs-type structure. The quaternary equiatomic ZrCuSiAs-type structure is very simple, with only eight atoms in the tetragonal cell. The dashed lines represent a unit cell.

Figure 15.2 shows the X-ray diffraction (XRD) pattern for a sample annealed at 1160°C. It is found that the peaks in theXRD pattern can be well indexed to the tetragonal ZrCuSiAs-type structure with $a = 0.3932$ nm and $c = 0.8490$ nm, except for some tiny peaks from the impurity phase SmOF. These lattice parameters are slightly smaller than the values of $a = 0.3940$ nm and $c = 0.8501$ nm for F-free SmFeAsO.

Magnetic characterization of the superconducting transitions under a magnetic field of 10 Oe for a sample annealed at 1160°C is shown in Fig. 15.3; data are shown for the zero-field cooled and field-cooled measurements. The magnetic onset for the superconducting transition is 41.8 K for the sample annealed at 1160°C, and 41.3 K for the sample annealed at 1200°C (not shown). The existence of the superconducting phase was unambiguously confirmed by the Meissner effect on cooling in a magnetic field. A superconducting volume

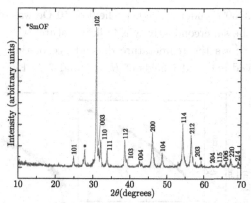

Fig. 15.2   X-ray diffraction pattern for a sample with nominal composition SmFeAs O$_{1-x}$F$_x$ ($x = 0.15$). The sample was annealed at 1160°C, and a tiny impurity phase SmOF is observed (stars denote peaks due to this impurity phase). The sample preparation process, except for annealing, was carried out in a glove box filled with a high-purity argon atmosphere. The samples were characterized at room temperature by X-ray diffraction using a Rigaku D/max-A X-ray diffractometer with Cu K$_\alpha$ radiation ($\lambda = 0.15418$ nm) in the $2\theta$ range of 10°−70° with steps of 0.02°.

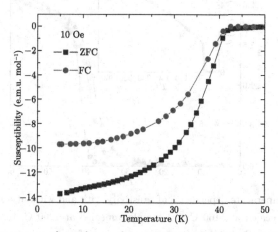

Fig. 15.3   Temperature dependence of magnetic susceptibility for a sample annealed at 1160°C. Data are shown for zero-field cooled (ZFC) and field-cooled (FC) measurements at 10 Oe. The susceptibility measurement was performed in an MPMS-7T system (Quantum Design).

fraction of about 50% under a magnetic field of 10 Oe was obtained at 5 K, indicating that the superconductivity is bulk in nature.

Figure 15.4 shows the temperature dependence of the resistivity under zero magnetic field and under fields of $H = 5$ and 7 T; Fig. 15.4(a) and (b)

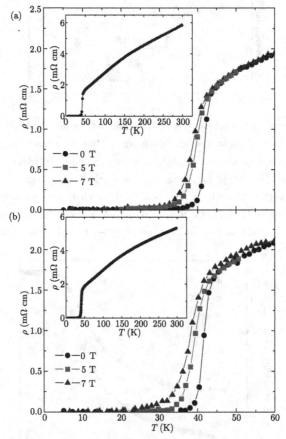

Fig. 15.4   Temperature dependence of resistivity with and without a magnetic field. (a) Sample annealed at 1160°C; (b) sample annealed at 1200°C. Insets, resistivity from 300 K to 5 K. Resistivity measurements were performed using an a.c. resistance bridge (Linear Research Inc., Model LR700) by the standard four-probe method. The transport properties were measured under magnetic fields of 5 and 7 T with an MPMS-7T system (Quantum Design).

show data for samples annealed at 1160°C and 1200°C, respectively. Under zero magnetic field, the onset transition and midpoint temperatures of the resistive transition are respectively 43 K and 41.7 K for a sample annealed at 1160°C; for a sample annealed at 1200°C, the values are respectively 43.7 K and 41.2 K. The 90-10% transition width are 2.5 K and 3 K for the samples annealed at 1160°C and 1200°C, respectively. It is found that the onset transition temperature in susceptibility coincides with the transition midpoint temperature in resistivity. An external magnetic field of 5 or 7 T makes the transition width broader, but the onset transition temperature is not sensitive to magnetic field, indicating that the upper critical field is very high for this superconductor. Therefore, this superconductor has potential applications due to its high transition temperature and high upper critical field.

Replacement of La by Sm leads to a large increase in $T_c$ from 26 K in $LaO_{1-x}F_xFeAs$ [3] to 43 K in $SmFeAsO_{1-x}F_x$ (this work). This suggests that it is possible to realize higher $T_c$ values in such layered oxypnictides. The observed $T_c$ of 43 K in $SmFeAsO_{1-x}F_x$ is higher than the theoretical value predicted from Bardeen-Cooper-Schrieffer (BCS) theory [1], and this provides a strong argument for considering layered oxypnictide superconductors as unconventional superconductors.

This work was supported by the Natural Science Foundation of China and by the Ministry of Science and Technology of China.

1] W. L. McMillan, Phys. Rev. **167**, 331 (1968).

2] J. Nagamatsu et al., Nature **410,** 63 (2001).

3] Y. Kamihara et al., J. Am. Chem. Soc. **130**, 3296 (2008).

4] P. Quebe et al., J. Alloys Comp. **302**, 70 (2000).

5] B. I. Zimmer et al., J. Alloys Comp. **229**, 238 (1995).

# 16

# Superconductivity at 55 K in Iron-Based F-Doped Layered Quaternary Compound Sm[O$_{1-x}$F$_x$]FeAs*

Z. A. Ren, W. Lu, J. Yang, W. Yi, X. L. Shen, Z. C. Li, G. C. Che, X. L. Dong, L. L. Sun, F. Zhou and Z. X. Zhao

*National Laboratory for Superconductivity, Institute of Physics and Beijing National Laboratory for Condensed Matter Physics, Chinese Academy of Sciences, PO Box 603, Beijing 100190*

We report the superconductivity in iron-based oxyarsenide Sm[O$_{1-x}$F$_x$]FeAs, with the onset resistivity transition temperature at 55.0 K and Meissner transition at 54.6 K. This compound has the same crystal structure as LaOFeAs with shrunk crystal lattices, and becomes the superconductor with the highest critical temperature among all materials besides copper oxides up to now.

The equiatomic transition metal quaternary oxypicnides have been studied for a long time [1,2]. Some Fe- and Ni-based oxypicnides have been found to be superconducting at low temperatures recently [3,4], and the very recent discovered F-doped La[O$_{1-x}$F$_x$]FeAs with a superconducting critical temperature $T_c$ = 26 K [5], is of great interest because of its higher $T_c$, layered structure and iron-containing character. The later experiments with the replacement of La by other rare earth elements, such as Sm, Ce, Pr, and Nd [6−9], has put this class to another high-$T_c$ family of superconductors above 50 K. All these arsenide (including phosphide) superconductors have been formed in the same tetragonal layered structure with the space group $P4/nmm$ that has an alternative stacked Fe-As layer and Re-O (Re = rare earth metals) layer, and $T_c$

---

*Reprinted from Z. A. Ren, W. Lu, J. Yang *et al.*, Chin. Phys. Lett. **25**, 2215 (2008).

is observed to be increased by the smaller rare earth substitution with shrunk
crystal lattice. In this Letter, we report our new results on the samarium-
arsenide $Sm[O_{1-x}F_x]FeAs$ synthesized under high pressure, with a resistivity
onset $T_c$ of 55.0 K, which is the highest among all materials besides copper
oxides up to now.

The superconducting $Sm[O_{1-x}F_x]FeAs$ samples were prepared by a high-
pressure (HP) synthesis method [9]. SmAs powder (pre-sintered) and As, Fe,
$Fe_2O_3$, $FeF_2$ powders (the purities of all starting chemicals are better than
99.99%) were mixed together according to the nominal stoichiometric ratio of
$Sm[O_{1-x}F_x]FeAs$, then ground thoroughly and pressed into small pellets. The
pellets were sealed in boron nitride crucibles and sintered in a high pressure
synthesis apparatus under the pressure of 6 GPa and temperature of 1250°C for
two hours. Compared with the common vacuum quartz tube synthesis method,
the HP method is more convenient and efficient for synthesize gas-releasing
compound with super-high pressure-seal. The structure of the samples was
characterized by powder X-ray diffraction (XRD) analysis on an MXP18A-
HF type diffractometer with Cu $K_\alpha$ radiation from 20° to 80° with steps of
0.01°.

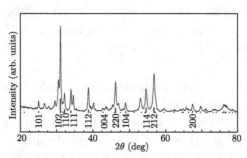

Fig. 16.1  X-ray powder diffraction pattern of the nominal $Sm[O_{0.9}F_{0.1}]FeAs$ compound; the vertical bars correspond to the calculated diffraction intensities.

The XRD patterns indicate that all the samples have a main phase of
LaOFeAs structure with some impurity phases, and the impurity phases have
been determined to be the known oxides, arsenides, and fluorides that were
formed by starting chemicals, which do not superconduct at the measuring
temperature. Here we note that because of the inevitable loss of fluorine

either by HP synthesis or ambient pressure synthesis, the real F-doped level is much smaller than the nominal one, and therefore the impurity phases always exist due to the unbalance of the stoichiometry for the nominal phase. The lattice parameters for all the samples are calculated by the least-square fit method with $|\delta 2\theta| < 0.01°$. For the undoped SmOFeAs, the lattice parameters $a = 3.933(5)$Å, $c = 8.495(4)$Å, while all superconducting samples have smaller lattices; for the nominal $Sm[O_{0.9}F_{0.1}]FeAs$, $a = 3.915(4)$Å and $c = 8.428(7)$Å. This result is different from the previous reported data where the crystal lattice was enlarged by F-doping [6] while consistent with all reports on other rare earth substitutions, and indicates the covalent character of the intra-layer chemical bonding due to the smaller covalent radius of fluorine than oxygen.

The dc resistivity was measured by the standard four-probe method. The results for the HP sample with a nominal composition of $Sm[O_{0.9}F_{0.1}]FeAs$ and an undoped SmOFeAs sample (sintered in sealed vacuum quartz tube) are shown in Fig. 16.2. The resistivity of SmOFeAs shows an anomaly at 150 K, which is similar to that of other ReOFeAs compounds that reported previously [5,7], and this anomaly was confirmed to be caused by the occurrence of spin-density-wave instability [12]. For $Sm[O_{0.9}F_{0.1}]FeAs$, the temperature of the onset resistivity transition was found to be at 55.0 K and the zero resistivity appeared at 52.6 K, which is higher than that of $Pr[O_{1-x}F_x]FeAs$ and $Nd[O_{1-x}F_x]FeAs$, and then becomes the highest among all superconducting

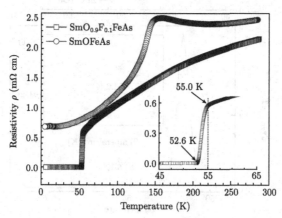

Fig. 16.2   Temperature dependence of resistivity for the undoped SmOFeAs and the $Sm[O_{0.9}F_{0.1}]FeAs$ superconductor.

materials besides copper oxides. As Sm has a smaller covalent radius comparing with La, Ce, Pr and Nd, the inner chemical pressure caused by the shrinkage of crystal lattice is thought of as an important factor to enhance $T_c$ [10], as proposed in a theoretical calculation in [11], where it was indicated that $T_c$ may be enhanced by the increase of hopping integral, which can be achieved by the shrinkage of the lattice.

The magnetization measurements were performed on a Quantum Design MPMS XL-1 system during warming cycle under fixed magnetic field after zero field cooling (ZFC) and field cooling (FC) process. The dc susceptibility data (measured under a magnetic field of 1 Oe) are shown in Fig. 16.3. The sharp magnetic transitions on the dc susceptibility curves indicate the good quality of this superconducting component. The onset diamagnetic transition

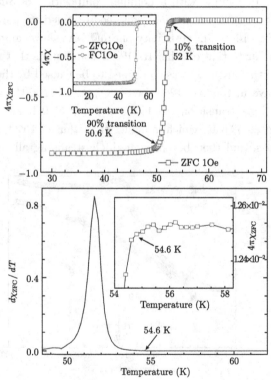

Fig. 16.3   Temperature dependence of the dc susceptibility and differential ZFC curve for the Sm[$O_{0.9}F_{0.1}$]FeAs superconductor.

starts at 54.6 K, and the 10% and 90% transitions on the ZFC curve are at 52 K and 50.6 K, respectively, with the middle of this Meissner transition at 51.5 K. For this class with much smaller rare earth substitution, higher $T_c$ might be expected, whereas samples with clear phase of the same structure are still absent.

We thank Mrs. Shun-Lian Jia for her kind help in resistivity measurements.

[1]  W. J. Zhu, Y. Z. Huang, C. Dong and Z. X. Zhao, Mater. Res. Bull. **29**, 143 (1994).

[2]  B. I. Zimmer, W. Jeitschko, J. H. Albering, R. Glaum and M. Reehuis, J. Alloys Comp. **229**, 238 (1995).

[3]  Y. Kamihara et al., J. Am. Chem. Soc. **128**, 10012 (2006).

[4]  T. Watanabe et al., Inorg. Chem. **46**, 7719 (2007).

[5]  Y. Kamihara, T. Watanabe, M. Hirano and H. Hosono, J. Am. Chem. Soc. **130**, 3296 (2008).

[6]  X. H. Chen, T. Wu, G. Wu, R. H. Liu, H. Chen and D. F. Fang, Nature **453**, 761 (2008) .

[7]  G. F. Chen, Z. Li, D. Wu, G. Li, W. Z. Hu, J. Dong, P. Zheng, J. L. Luo and N. L. Wang, Phys. Rev. Lett. **100**, 247002 (2008).

[8]  Z. A. Ren, J. Yang, W. Lu, W. Yi, G. C. Che, X. L. Dong, L. L. Sun and Z. X. Zhao, Materials Research Innovations **12**, 105 (2008).

[9]  Z. A. Ren et al., Europhysics Letters **82,** 57002 (2008).

[10]  W. Lu, J. Yang, X. L. Dong, Z. A. Ren, G. C. Che and Z. X. Zhao, New J. Phys. **10**, 063026 (2008).

[11]  Q. Han, Y. Chen and Z. D. Wang, Europhys. Lett. **82** 3707 (2008).

[12]  M. A. McGuire et al., arXiv:cond-mat/ 0804.0796.

# 17

# Anomalously Large Gap Anisotropy in the a-b Plane of $Bi_2Sr_2CaCu_2O_{8+\delta}$*

Z. X. Shen[1,2], D. S. Dessau[1,2], B. O. Wells[1,2,a], D. M. King[2], W. E. Spicer[2], A. J. Arko[3], D. Marshall[2], L. W. Lombardo[1], A. Kapitulnik[1], P. Dickinson[1], S. Doniach[1], J. Dicarlo[1,2], A. G. Loeser[1,2] and C. H. Park[1,2]

[1] *Department of Applied Physics, Stanford University, Stanford, California 94305*
[2] *Solid State Electronics Laboratory and Stanford Synchrotron Radiation Laboratory, Stanford University, Stanford, California 94305-4055*
[3] *Los Alamos National Laboratory, Los Alamos, New Mexico 87545*

Superconducting gap anisotropy at least an order of magnitude larger than that of the conventional superconductors has been observed in the a-b plane of $Bi_2Sr_2CaCu_2O_{8+\delta}$ in angle-resolved photoemission spectroscopy. For samples with $T_c$, of 88 K, the gap size reaches a maximum of approximately 20 meV along the Cu-O bond direction, and a minimum of much smaller or vanishing magnitude 45° away. The experimental data are discussed within the context of various theoretical models. In particular, a detailed comparison with what is expected from a superconductor with a $d_{x^2-y^2}$ order parameter is carried out, yielding a consistent picture.

A key to understanding the mechanism of high-$T_c$, superconductivity is the symmetry of the superconducting order parameter. The conventional BCS theory has an s-wave order parameter, reflecting the spherically symmetric nature of the pair wave function. For the high-$T_c$ cuprates, theoretical analysis of the crucial $CuO_2$ plane with consideration of the strong on-site Coulomb interaction leads to other symmetries of the order parameter [1–10]. In par-

---

ticular, pairing theories based on the Hubbard model or its derivatives lead to a $d$-wave order parameter or a mixed symmetry order parameter with a strong $d$-wave component [1–10]. Very recently, stimulated by the NMR data, this issue of the order parameter has once again attracted great attention in the field [7,8].

Angle-resolved photoemission (ARPES) from $Bi_2Sr_2CaCu_2O_{8+\delta}$ (Bi2212) has played an important role in helping us understand the cuprate super-conductors. For Bi2212, ARPES is sensitive to both the normal-state Fermi surface and the superconducting gap [11–15]. In fact, its ability to measure the superconducting gap as a function of crystal momentum is currently a unique capability, providing an opportunity to probe the symmetry of the order parameter. Although the superconducting gap as revealed by ARPES only reflects the magnitude of the order parameter, it still provides important constraints for theoretical models. In the past, conflicting results have been published [16,17]. An earlier study showed that the superconducting gap did not vary in $k$ space, and therefore provided support for an s-wave order parameter [16]. Recently, we reported preliminary results showing sizable su-perconducting gap anisotropy in the $a$-$b$ plane [15,17], in contrast to the earlier report.

In this paper, we report more comprehensive data from our recent study of Bi2212 showing a gap anisotropy in the $a$-$b$ plane that is at least an order of magnitude larger than that of the conventional superconductors. In addition, we have observed spectra that are indicative of a node in the gap within the experimental limitation. We also discuss in detail the difficulties in the data acquisition and interpretation as a result of finite energy resolution, limited sample lifetime, and other material problems. Next, we address the ARPES results in the context of various theoretical models. In particular, we have compared our data with what is expected from the $d_{x^2-y^2}$ order parameter $\Delta(k) \sim \cos k_x a - \cos k_y a$, yielding a very consistent picture.

The qualitative picture of the strong gap anisotropy in $k$ space has been reproduced in many samples during six different experimental runs using the two chambers. For data in this paper, sample 1 ($T_c \sim 78$ K) was measured in a VSW system using a He discharge lamp under conditions similar to those reported before [15,17]. Samples 2 and 3 ($T_c \sim 88$ K) were measured in another VSW chamber at the beam line 5 of SSRL. The combined energy resolution of the analyzer and the monochromator is about 30 meV. Unless otherwise stated, the samples were cleaved and measured at 35 K. All samples reported in this paper have a superconducting transition width of 2 K or smaller as determined by Meissner shielding. The samples were introduced into the photo emission chamber through a load lock without baking. This procedure is necessary to ensure the sharp superconducing transition, as verified by $T_c$ determinations of several samples before and after the ARPES experiments. The nominal chamber pressure during the measurement was $1 \times 10^{-10}$ torr.

Figure 17.1 presents normal (open squares) and superconducting state (solid circles) ARPES spectra from sample 1. The $\boldsymbol{k}$ space locations were chosen so that the normal-state peak is at the Fermi level, and the midpoint of the leading edge in the normal state coincides with the Fermi level at both points $A$ and $B$. Very clear spectral changes are observed at $A$ as the sample

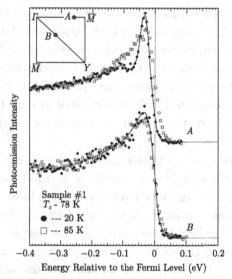

Fig. 17.1   High resolution photoemission spectra from sample 1 recorded at $\boldsymbol{k}$ space locations $A$ and $B$, as illustrated in the inset. The spectra at $B$ were measured before those at $A$. The spectral changes above and below $T_c$ are caused by the opening of the superconducting gap. The change at $A$ is quite visible, yielding a larger gap. The change at $B$ is hardly visible, suggesting a very small or null gap.

is cooled below $T_c$. The leading edge of the superconducting spectrum is pulled back to higher binding energy, reflecting the opening of the superconducting energy gap. At the same time, a "pileup" and a "dip" near $-80$ meV appear in the data [15,18]. At $B$, only minor changes with temperature are observed, indicating that the gap is undetectable within experimental uncertainty. This striking difference at the two $k$ space locations indicates that the superconducting gap is very anisotropic.

To quantify the gap anisotropy is an important but difficult task. There are four aspects to the experimental difficulties. First, there is our finite energy resolution which limits the precision and the accuracy of our measurements. Next, energy calibration fluctuations of the spectrometer mainly caused by a drift in the electronics add uncertainty to the Fermi energy location. For data in this paper, the uncertainty is about 1 meV for sample 1, and $2-3$ meV for samples 2 and 3. Third, the sample surface flatness and the finite angular resolution limit the momentum resolution of our experiment. The surface flatness is sample dependent, causing a scatter in the experimental data. This remains a technical problem despite our best efforts at selecting the best possible crystals and refining the cleaving technique. Our characterizations show that the surfaces of samples 1 and 3 are flatter that sample 2. Finally, there is the time dependence of the data which will be illustrated in Fig. 17.2. ARPES is a surface-sensitive technique so the spectra may change as the sample surface ages.

Figure 17.2 shows spectra from sample 2 at the $\overline{M}$ point recorded at different times after the sample was cleaved. With time, the superconducting quasiparticle peak shifts to lower binding energy, indicating that the superconducting gap becomes smaller. At the same time, the dip structure at $-80$ meV also becomes smaller. Importantly, the clean sample surface can be regenerated by warming the sample up to room temperature. The spectrum marked with "reg + 0:25" was taken 25 min after the surface had been regenerated and cooled down again. Both the larger superconducting gap size and the $-80$ meV dip are reproduced in the data. This provides additional support for the suggestion that the dip is an intrinsic superconducting property [15,18]. The ability to regenerate the samples strongly suggests that the changes are due to physisorption of the residual gases onto the sample surface at low temperature.

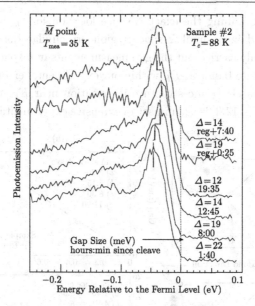

Fig. 17.2    Spectra from sample 2 recorded at $\overline{M}$ at different times after the sample was cleaved and kept at low temperature. As the sample ages, the superconducting gap becomes smaller. The numbers marked are the gap size and its aging time after the cleave. A decrease of the intensity of the −80 meV dip is clearly visible. The clean sample surface can be regenerated by warming up the sample to room temperature, and then cooling down again.

The theoretical difficulty in quantifying the gap anisotropy is mainly caused by the fact that we do not have an adequate theory to describe the angle-resolved photoemission line shape of either the normal or the superconducting state. For simplicity in determining the gap size without any specific fittings, we chose to call the gap the energy position of the midpoint of the leading edge of the superconducting state spectrum. With this criterion, the spectrum at point $A$ in Fig. 17.1 has a gap of 12 meV ($2\Delta/k_BT_c = 3.6$) and the spectrum at point $B$ gives a gap of 0 to 0.5 meV. In the limit of perfect angular resolution, this method requires that the normal-state band ($\varepsilon_{\boldsymbol{k}}$) be at the Fermi level. With finite angular resolution, the uncertainty in the gap size caused by the normal-state band being slightly off $\varepsilon_{\boldsymbol{k}} = 0$ is reduced. The qualitative picture of the observed gap anisotropy is not sensitive to the experimental

difficulty in determining the exact $\boldsymbol{k}$ space location for $\varepsilon_{\boldsymbol{k}} = 0$ because the
gap values found in a small $\boldsymbol{k}$ space region are similar (as can be seen in
Fig. 17.3). We only carried out the gap measurements in $\boldsymbol{k}$ space regions where
we see normal-state bands at $E_F$ within our experimental error (shaded area).
The normal-state electronic structure information near $E_F$, depicted by the
shaded area in Fig. 17.3, is a summary of extensive experimental measurements

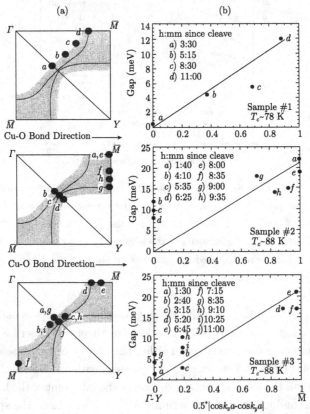

Fig. 17.3   (a) The Brillouin zone locations where the gap is measured. The shaded
areas are $\boldsymbol{k}$ space locations where we found bands very close to $E_F$. The solid lines
represent the Fermi surface due to the two CuO$_2$ planes only [20]. (b) Gap size vs
$0.5|\cos k_x a - \cos k_y a|$. The straight lines are predictions of the $d_{x^2-y^2}$ order parameter.
Sample 1 has higher oxygen content and lower $T_c$. This could account for the smaller
$\Delta$, near $\overline{M}$.

that will be published in the future [19]. The experimental results are consistent with an earlier study where only the data along $\Gamma$-$Y$ line were reported [11]. Furthermore, a very flat band at $E_F$ is observed for a large $\boldsymbol{k}$ space region near $\overline{M}$, which is similar to results from YBCO at comparable $\boldsymbol{k}$ space locations [21].

Figure 17.3 displays the measured $\boldsymbol{k}$ dependence of the gap from three samples. In order to account for the time effects, the spectra from the different samples were taken using different time sequences. We limit ourselves to spectra recorded within 12 h after the sample was cleaved. Each spectrum is labeled with a letter, with the order of measurements corresponding to the alphabetic progression. The $\boldsymbol{k}$ space location of the measurement is shown in panel (a), and the magnitude of the gap for each measurement is displayed on the vertical axes of the graphs in panel (b). The horizontal axis of panel (b) is $0.5|\cos k_x a - \cos k_y a|$, allowing a direct comparison with the $d$-wave theory, as will be discussed later. This function is zero along the $\Gamma$-$Y$ line and increases monotonically away from the line and approaches its maximum at $\overline{M}$. Despite the potential problem with time effects, the data in Fig. 17.3 clearly suggest that the gap is smallest along the $\Gamma$-$Y$ direction in the Brillouin zone, and it gets bigger as one moves away from the $\Gamma$-$Y$ line. This is true for all samples, and is independent of the measuring time sequence. The magnitude of the measured gap anisotropy is very large, ranging roughly between a factor of 2 and 10, depending upon the sample. This is 1 to 2 orders of magnitude larger than what has been observed in conventional superconductors [22].

Although the qualitative trend of gap variation in $\boldsymbol{k}$ space is very reproducible, the gap size shows some scatter along the $\Gamma$-$Y$ line. In Fig. 17.3(b), samples 1 and 3 show a very small gap along $\Gamma$-$Y$ (where $|\cos k_x a - \cos k_y a| = 0$), while sample 2 shows a significantly larger gap in this region. A hint for the explanation of the scatter in the data can be found in the time dependence of the spectra from sample 3 taken along the $\Gamma$-$Y$ direction and the sample quality as indicated by the laser reflection. The earliest spectrum from sample 3 was taken 1.5 h after the cleave and showed a small (1.5 meV) gap. Approximately 7 h later, the gap at that point was observed to grow to 6.5 meV, after which time it again shrank to near zero after the sample surface was regenerated. The direction of this time dependence is, interestingly, *opposite* to that observed in the region around $\overline{M}$ where the gap is large. We can make sense of all this data by the following hypothesis. The intrinsic gap along $\Gamma$-$Y$ is very small (or zero). Both sample 2 and the "aged" spectra of sample 3 had significantly poorer $\boldsymbol{k}$ resolution (the physisorbed layer of gas causing the extra scattering for the "aged" sample), producing a spectrum with a significant (but still small) gap by averaging in region of $\boldsymbol{k}$ space which had a nonzero gap. No other argument that we are aware of can account for the time and sample dependence of the data as well. This analysis is consistent with the fact that samples 1 and 3 have sharper laser reflections. Given the analysis, we believe that our body of data is consistent with a very small or zero gap along $\Gamma$-$Y$. In fact, it is fair to state that the spectra at $B$ in Fig. 17.1 suggest that the superconducting gap along $\Gamma$-$Y$ is zero within our experimental uncertainty.

The observed superconducting gap variation can be explained by considering only the Cu-O states. Although there is some hybridization of Bi-O states near $E_F$ around $\overline{M}$ [14], one would not expect a larger gap from the Bi-O states since the superconducting properties originate in the CuO$_2$ layers. In addition, we see a similar gap anisotropy in samples with different oxygen content, which we know the Bi-O states are very sensitive to [14]. For example, compared to samples 2 and 3, sample 1 has a higher oxygen content and somewhat lower $T_c$, resulting in a large reduction in the gap value at $\overline{M}$; however, the general trend of the gap variation is the same.

Several theoretical models can qualitatively explain the observed gap anisotropy. First, the mechanism of superconductivity based on interlayer coupling attributes the small or vanishing gap along the $\Gamma$-$Y$ direction to the vanishing coupling matrix element between the two CuO$_2$ planes [23]. Second, the mechanism based on the van Hove scenario may be able to attribute the larger gap along $\Gamma$-$\overline{M}$ as having been stabilized by the van Hove singularity [24,25]. Finally, any theoretical model with the $d_{x^2-y^2}$ order parameter can also explain our data. We will concentrate on the last scenario because the first two have not been fully developed to give a more detailed comparison with our experiment at this stage. The $d_{x^2-y^2}$ order parameter $\Delta(k) \sim \cos k_x a - \cos k_y a$ will yield a gap that is proportional to $|\cos k_x a - \cos k_y a|$, as shown by the straight lines in Fig. 17.3(b). The experimental data agree with the $d$-wave picture well. The fact that the scatter in the measured gap size is larger along $\Gamma$-$Y$ than along $\Gamma$-$\overline{M}$ is also consistent with the $d$-wave scenario because $|\cos k_x a - \cos k_y a|$ is most sensitive to the $k$ averaging along the $\Gamma$-$Y$

direction. Furthermore, the data in Fig. 17.1 and the discussion of Fig. 17.3 are suggestive of the expected $d$-wave node within the experimental uncertainty. In the same context, we have also considered order parameters with other symmetries such as extended $s$-wave or $s+id$. Our data are qualitatively incompatible with the extended $s$-wave scenario, having the gap proportional to $|\cos k_x a + \cos k_y a|$. The present state of the data does not allow us to distinguish pure $d_{x^2-y^2}$ symmetry from a mixed symmetry of $s+id$ because we cannot definitely establish the existence of the node line. However, since the overall agreement of the data to the form $|\cos k_x a - \cos k_y a|$ is good, the data suggest a strong $d_{x^2-y^2}$ component even if a node does not exist.

The above interpretation of our data is another piece of circumstantial evidence for $d$-wave superconductivity. On the one hand, it is consistent with the $d$-wave interpretation of the observed anisotropy in the NMR relaxation rate between copper and oxygen [7,8]. On the other hand, it is not consistent with the temperature dependence of the penetration depth of $Nd_{1.85}Ce_{0.15}CuO_4$ [26].

To conclude, we have observed a superconducting gap anisotropy in the $a$-$b$ plane of Bi2212 that is at least an order of magnitude larger than that of the conventional superconductors. All aspects of the data (the anomalously large gap anisotropy, the specific $k$ dependence of the anisotropy, and the possible presence of the node along $\Gamma$-$Y$) can be well understood if we assume Bi2212 is a superconductor with the $d_{x^2-y^2}$ order parameter.

The data presented here were obtained from the Stanford Synchrotron Radiation Laboratory (SSRL), which is operated by the DOE Office of Basic Energy Sciences, Division of Chemical Sciences. The Office's Division of Materials Science has provided support for this research. The Stanford work was also supported by NSF Grants No. DMR8913478 and No. DMR9121288, and the NSF grant through the Center of Material Research. Beam line 5 of SSRL was built with DARPA, ONR, AFOSR, AOR, DOE, and NSF support. The University of Maryland work was supported by NSF Grant No. DMR9115384.

[1] D. J. Scalapino *et al.*, Phys. Rev. B **34**, 8190 (1986); N. E. Bickers *et al.*, Int. J. Mod. Phys. B **1**, 687 (1987); N. E. Bickers *et al.* , Phys. Rev. Lett. **62**, 961 (1989).

[2] A. Ruckenstein *et al.*, Phys. Rev. B **36**, 857 (1987).

[3] G. Kotliar and J. Liu, Phys. Rev. B **38**, 5142 (1988).

[4] J. R. Schrieffer, X. G. Wen and S. C. Zhang, Phys. Rev. Lett. **60**, 944 (1988).

[5] M. Inui *et al.*, Phys. Rev. B **37**, 2320 (1988).

[6] G. J. Chen, R. Joynt and F. C. Zhang, Phys. Rev. B **42**, 2662 (1990).

[7] P. Monthoux *et al.*, Phys. Rev. Lett. **67**, 3448 (1991);**69**, 961 (1992).

[8] N. Bulut and D. J. Scalapino, Phys. Rev. Lett. **68**, 706 (1992).

[9] D. Rokhsar (to be published).

[10] P. H. Dickinson and S. Doniach (to be published).

[11] C. G. Olson *et al.*, Science **245**, 731 (1989).

[12] J. M. Imer *et al.*, Phys. Rev. Lett. **62**, 336 (1989).

[13] R. Manzke *et al.*, Europhys. Lett. **9**, 477 (1989).

[14] B. O. Wells *et al.*, Phys. Rev. Lett. **65**, 3056 (1990).

[15] D. S. Dessau *et al.*, Phys. Rev. Lett. **66**, 2160 (1991).

[16] C. G. Olson *et al.*, Solid State Commun. **76**, 411 (1990).

[17] B. O. Wells *et al.*, Phys. Rev. B **46**, 11830 (1992).

[18] D. S. Dessau *et al.*, Phys. Rev. B **45**, 5095 (1992).

[19] D. S. Dessau, Ph. D. thesis, Stanford University, 1992.

[20] S. Massida, J. Yu and A. J. Freeman, Physica **152C**, 251 (1988).

[21] R. Liu *et al.*, Phys. Rev. B **45**, 5614 (1992); J. G. Tobin *et al.*, *ibid.* **45**, 5563 (1992).

[22] A. G. Sheplev, Usp. Fiz. Nauk, **96**, 217 (1969) [Sov. Phys. Usp. **11**, 690 (1969)]; here we do not include the heavy fermion superconductors in our comparison.

[23] P. W. Anderson (private communication).

[24] R. S. Markiewicz, Physica **153-155C**, 1181 (1988).

[25] D. M. Newns *et al.* , Comments Condens. Matter Phys. B **15**, 273 (1992).

[26] D. H. Wu *et al.* , Phys. Rev. Lett. **70**, 85 (1993).

# 18
# Identification of a New Form of Electron Coupling in the Bi$_2$Sr$_2$CaCu$_2$O$_8$ Superconductor by Laser-Based Angle-Resolved Photoemission Spectroscopy*

W. T. Zhang[1], G. D. Liu[1], L. Zhao[1], H. Y. Liu[1], J. Q. Meng[1], X. L. Dong[1], W. Lu[1], J. S. Wen[2], Z. J. Xu[2], G. D. Gu[2], T. Sasagawa[3], G. L. Wang[4], Y. Zhu[5], H. B. Zhang[4], Y. Zhou[4], X. Y. Wang[5], Z. X. Zhao[1], C. T. Chen[5], Z. Y. Xu[4] and X. J. Zhou[1]

[1] *National Laboratory for Superconductivity, Beijing National Laboratory for Condensed Matter Physics, Institute of Physics, Chinese Academy of Sciences, Beijing 100080, China*
[2] *Condensed Matter Physics and Materials Science Department, Brookhaven National Laboratory, Upton, New York 11973, USA*
[3] *Materials and Structures Laboratory, Tokyo Institute of Technology, Yokohama Kanagawa, Japan*
[4] *Laboratory for Optics, Beijing National Laboratory for Condensed Matter Physics, Institute of Physics, Chinese Academy of Sciences, Beijing 100080, China*
[5] *Technical Institute of Physics and Chemistry, Chinese Academy of Sciences, Beijing 100080, China*

Laser-based angle-resolved photoemission measurements with superhigh resolution have been carried out on an optimally doped Bi$_2$Sr$_2$CaCu$_2$O$_8$ high temperature superconductor. New high energy features at ~115 meV and ~150 meV, in addition to the prominent $\sim$70 meV one, are found to develop in the nodal electron self-energy in the superconducting state. These high energy features, which cannot be attributed to electron coupling with single phonon or magnetic resonance mode, point to the existence of a new form of electron coupling in high temperature superconductors.

The physical properties of materials are dictated by the microscopic electron dynamics that relies on the many-body effects, i.e., the interactions of electrons with other excitations, such as phonons, magnons, and so on. How to detect and disentangle these many-body effects is critical to understanding the macroscopic physical properties and the superconductivity mechanism in high temperature superconductors. Angle-resolved photoemission spectroscopy (ARPES), as a powerful tool in probing many-body effects [1], has revealed clear evidence of electron coupling with low-energy collective excitations (bosons) at an energy scale of $\sim$70 meV [2−7] in the nodal region and $\sim$40 meV near the antinodal region [4,8−10] although the nature of the bosonic modes remains under debate as to whether it is phonon [5,6,10] or magnetic resonance mode [3,4,7,8,11]. Recently, another high energy feature has been identified in dispersion at 300−400 meV, but its origin remains unclear as to whether this can be attributed to a many-body effect [12].

In this Letter we report an identification of a new form of electron coupling in high temperature superconductors by taking advantage of a superhigh resolution vacuum ultraviolet (VUV) laser-based ARPES technique [13]. New features at energy scales of $\sim$115 meV and $\sim$150 meV are revealed in the electron self-energy in the $Bi_2Sr_2CaCu_2O_8$ (Bi2212) superconductor in the superconducting state. These features cannot be attributed to electron coupling with single phonon mode or magnetic resonance mode. They point to a possibility of electron coupling with some high energy excitations in high temperature superconductors.

The angle-resolved photoemission measurements have been carried out on our newly developed VUV laser-based angle-resolved photoemission system [13]. The photon energy of the laser is 6.994 eV with a bandwidth of 0.26 meV. The energy resolution of the electron energy analyzer (Scienta R4000) is set at 0.5 meV, giving rise to an overall energy resolution of 0.56 meV, which is significantly improved from 10−15 meV from regular synchrotron radiation systems [2−7]. The angular resolution is $\sim$0.3°, corresponding to a momentum resolution $\sim$0.004 Å$^{-1}$ at the photon energy of 6.994 eV, more than twice improved from 0.009 Å$^{-1}$ at a regular photon energy of 21.2 eV for the same angular resolution. The photon flux is adjusted between $10^{13}$ and $10^{14}$ photons/second. The optimally doped Bi2212 single crystals with a superconducting transition temperature $T_c = 91$ K were cleaved *in situ* in vacuum with a base pressure better than $5 \times 10^{-11}$ Torr.

Figure 18.1(a) shows the raw data of photoelectron intensity as a function of energy and momentum for an optimally doped Bi2212 superconductor ($T_c = 91$ K) measured along the $\Gamma(0,0) - Y(\pi,\pi)$ nodal direction at a temperature of 17 K. By fitting momentum distribution curves (MDCs), the dispersion [Fig. 18.1(b)] and MDC width [inset of Fig. 18.1(b)] are quantitatively extracted from Fig. 18.1(a). One can see an obvious kink in dispersion near 70 meV [Figs. 18.1(a) and 18.1(b)] and a drop in the MDC width [inset of Fig. 18.1(b)], similar to those reported before [2–7] but with much improved clarity. It

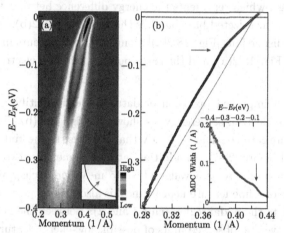

Fig. 18.1   Electron dynamics of optimally doped Bi2212 ($T_c = 91$ K) measured along the $\Gamma(0,0) - Y(\pi,\pi)$ nodal direction at 17 K. (a) Raw image showing photoelectron intensity (represented by false color) as a function of energy and momentum. The inset shows the location of the momentum cut in the Brillouin zone; (b) nodal dispersion extracted from (a) by fitting MDCs. The dotted line connecting the two energy positions in the dispersion at the Fermi energy and $-0.4$ eV is an empirical bare band for extracting the effective real part of self-energy in Fig. 18.2(b). The inset shows the corresponding MDC width (FWHM).

is generally agreed that this 70 meV feature originates from a coupling of electrons with a collective boson mode. When coming to the nature of the boson mode, it remains under debate whether it is phonon [5,6] or magnetic resonance mode [3,4,7].

The real part of the electron self-energy can be extracted from the dispersion given that the bare band dispersion is known which can be determined in a number of ways but still without consensus [14,15]. To identify fine features in the electron self-energy and study their relative change with temperature, it is reasonable to assume a featureless bare band for the nodal dispersion within a small energy window near the Fermi energy. In this case, the fine features manifest themselves either as peaks or curvature changes in the "effective self-energy" [15]. As shown in Fig. 18.1(b), we choose here a straight line connecting two energy positions in the dispersion at the Fermi energy and $-0.4$ eV as the empirical bare band. The resultant effective real part of electron self-energy, which represents the energy difference between the measured dispersion and the selected bare band, is shown in Fig. 18.2(b). Also shown in Fig. 18.2 are dispersions [Fig. 18.2(a)] along several other cuts in the Brillouin zone [inset of Fig. 18.2(a)] and the corresponding effective electron self-energy [Fig. 18.2(b)].

With much improved precision of data, one can identify clearly several features in the electron self-energy, as shown in Fig. 18.2(b). The most pronounced feature is the peak at $\sim$70 meV that gives rise to the kink in dispersion seen here and before [2−7]. In addition, at higher energies, two new features can be identified clearly as a valley at $\sim$115 meV and a cusp at $\sim$150 meV. The signature of a fine feature near 94 meV is also visible, particularly for the two cuts close to the nodal region (cuts 1 and 2). Between the Fermi level and 70 meV, we have also observed hints of possible low-energy features which are, however, very subtle and need further measurements to pin them down. We note that the bare band selection has little effect on the identification of these fine features and their energy position as we have checked by trying other straight lines as empirical bare bands. Particularly, the two new features at 115 meV and 150 meV together with the 70 meV peak, are robust (as also shown in Fig. 18.4 below) and persist in a rather large momentum space near the nodal region.

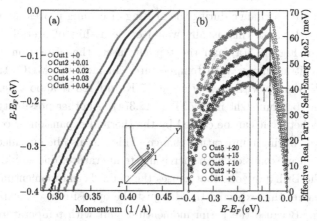

Fig. 18.2  Momentum dependence of dispersions (a) and corresponding effective real part of electron self-energy (b). The inset of (a) shows the location of momentum cuts in the Brillouin zone. The effective self-energy in (b) is obtained by taking the straight line connecting two positions in dispersion at the Fermi energy and $-0.4$ eV as a bare band. The arrows in (b) mark fine features at $\sim 70$ meV, 115 meV, and 150 meV, and a possible feature near 94 meV. For clarity, curves in (a) are offset along the momentum axis, while curves in (b) are offset along the vertical axis; the offset values are given in the legends.

The nodal electron dynamics undergoes a dramatic evolution with temperature and superconducting transition, as indicated by the temperature dependence of the nodal dispersion [Fig. 18.3(a)] and scattering rate [Fig. 18.3(b)]. In Fig. 18.3(a), a quantitative momentum variation with temperature at four typical energy positions (the Fermi level, $-0.07$ eV, $-0.2$ eV, and $-0.3$ eV) is plotted in the upperleft inset, and some representative MDCs for these four energies at 17 K and 128 K are plotted in the bottom-right inset. Over the temperature range of the measurement, the dispersion change with temperature spreads over an energy range of $0-300$ meV within which the dispersion renormalization gets stronger with decreasing temperature.

An unexpected finding is the Fermi momentum shift with temperature [Fig. 18.3(a) and top-left inset], which increases with increasing temperature in the superconducting state below $T_c = 91$ K and then becomes nearly flat above $T_c$. The magnitude of the shift is small, on the order of 0.003 Å$^{-1}$, and the change is monotonic. We first checked whether this could be caused by a sample orientation change during the heating or cooling process and feel that it is unlikely because this usually would cause a shift of overall dispersion. As shown in Fig. 18.3(a) and the top-left inset, the dispersion above 300 meV shows little change with temperature. In fact, the MDCs themselves at 300 meV overlap with each other at 17 K and 128 K almost perfectly, as shown in the bottom-right inset of Fig. 18.3(a). Another possibility we have checked is whether it can be caused by the thermal expansion or contraction of the sample during temperature change. This can also be excluded because with the lattice expansion with increasing temperature, one would expect a shrink of the Brillouin zone that causes the nodal Fermi momentum move to a smaller value; this expected trend is just the opposite to our experimental observation. Because the Fermi momentum shift with temperature is quite unusual and has not been reported before, we have repeated the measurement

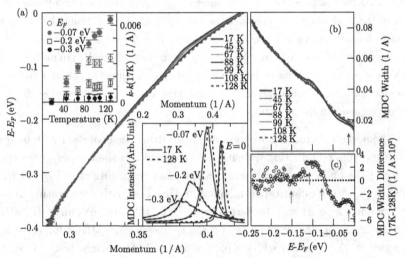

Fig. 18.3   Temperature dependence of the nodal MDC dispersion (a) and MDC width (b). The top-left inset of (a) plots the momentum value as a function of temperature at four typical energies: $E_F$ [empty (red) circle], $-0.07$ eV [solid (green) circle], $-0.2$ eV [empty (blue) square], and $-0.3$ eV [solid (black) circle] obtained by averaging over the $\pm 10$ meV energy range. The bottom-right inset of (a) shows MDCs at these four energies measured at 17 K (solid line) and 128 K (dashed line). (c) The difference of the MDC width between two temperatures at 17 K and 128 K. The arrows in (b) and (c) mark possible features showing up in the scattering rate.

and reproduced the similar observation. Therefore, to the best of our efforts, we think this is likely an intrinsic effect and its first observation is a result of much improved instrumental precision. Further work needs to be done to pin down this effect and understand the underlying physical origin as it suggests either a chemical potential shift or Fermi surface topology change upon entering the superconducting state.

The scattering rate shows a strong variation with temperature at the low-energy range within 0 to −0.2 eV, as shown in Fig. 18.3(b). Interestingly, the temperature dependence is not monotonic but depends on the binding energy. Between the Fermi level and −0.07 eV, the scattering rate decreases with decreasing temperature, while it increases with decreasing temperature between −0.07 and −0.15 eV. This gives rise to an "overshoot" region extending to ∼ −0.1 eV at low temperature. The change of the scattering rate with temperature can also be clearly seen in the difference between the normal and superconducting states, as plotted in Fig. 18.3(c) which depicts the difference between 17 K and 128 K data. Here the difference between −0.07 eV and −0.15 eV is positive while it becomes negative between $E_F$ and −0.07 eV.

The temperature dependence of the effective real part of self-energy (Fig. 18.4) indicates that the new high energy features at 115 meV and 150 meV are developed in the superconducting state. Figure 18.4(a) shows the effective real part of self-energy at various temperatures obtained from the dispersions [Fig. 18.3(a)] by selecting a common empirical bare band, i.e., a straight line connecting the Fermi energy and −0.4 eV in the dispersion at 128 K. Figure 18.4(b) shows the net temperature change of electron self-energy with respect to the normal state data at 128 K, thus avoiding any ambiguity from bare band selection. It is clear in both cases that a dramatic change of electron self-energy occurs in the superconducting state, with a sharpening of the ∼70 meV feature, together with the emerging and growing of the ∼115 meV and ∼150 meV features. These observations are consistent with the scattering rate data where one can see the emergence of similar characteristic energy scales at ∼150 meV, ∼115 meV, and ∼70 meV in the superconducting state [Fig. 18.3(c)].

The identification of high energy features at 115 meV and 150 meV points to a new form of electron coupling in high temperature superconductors. These are qualitatively different from the ∼70 meV feature which is attributed to coupling of electrons with some collective modes, and such modes with comparable energy scales are available in high temperature superconductors either as phonon [5,6] or magnetic resonance mode [3,4,7]. Because the energy scale of these two new features is higher than the maximum energy of phonons (~90 meV) [16] and the magnetic resonance mode (42 meV in optimally doped Bi2212) [17], they cannot be due to electron coupling with a single phonon or magnetic resonance mode. Although the effect of the electron-boson coupling for a low-energy mode can extend to high energy in the electron self-energy, it will not generate any new features with clear curvature change, as evidenced from simulations of electron-phonon coupling using both Debye and Einstein models and confirmed in canonical electron-phonon coupling systems [18].

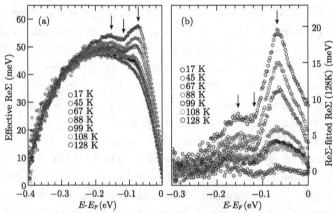

Fig. 18.4    (a) Temperature dependence of the effective real part of electron self-energy extracted from dispersions in Fig. 18.3(a) by taking a straight line as a bare band connecting two points at $E_F$ and $-0.4$ eV on the dispersion at 128 K. (b) Temperature dependence of the difference between the measured self-energy in (a) and the fitted one for 128 K [solid black line in (a)]. The 128 K self-energy is fitted by polynomials and used to reduce statistical errors.

There are a couple of possibilities that may give rise to high energy features in the electron coupling. The first is the mode energy shift due to the opening of superconducting gap: the original mode position in the normal state is expected to be shifted upward by an amount on the order of the superconducting gap upon entering the superconducting state [19]. In the optimally doped Bi2212, with the maximum $d$-wave superconducting gap at $\sim$35 meV, one might expect that the original $\sim$70 meV mode be shifted to a higher energy around 105 meV. However, this scenario is difficult to explain the existence of another 150 meV feature, the feature at 110 meV being a valley instead of a peak and the remaining strong $\sim$70 meV feature. Another possibility is the electron coupling with multiple phonons. Usually this effect is expected to be much weaker [20] although, in principle, the possibility cannot be totally excluded. More theoretical work is needed to verify whether such a multiphonon process can produce clear features at high energy and whether such an effect is enhanced at low temperature. The third possibility is electron coupling with excitations that are already present at such high energy scales. The emergence and evolution of the 115 meV and 150 meV features in the electron self-energy is probably due to the redistribution of the underlying spectral function of the high energy excitations with temperature and superconducting transition. One candidate of such high energy excitations in high temperature superconductors seems to be naturally related to the spin fluctuation observed by neutron scattering, which covers a large energy range up to 200 meV and exhibits strong temperature dependence [21]. The signature of such high energy coupling is also proposed from optical measurements [22]. Further theoretical work to investigate the effect of such high energy spin excitations on electron dynamics will help in clarifying such a scenario.

In conclusion, by performing high precision ARPES measurements on Bi2212, we have revealed new features at 115 meV and 150 meV in the electron self-energy developed in the superconducting state. They cannot be attributed to electron coupling with either the single phonon or magnetic resonance mode, but point to the existence of a new form of electron coupling in high temperature superconductors. We hope this observation will stimulate further theoretical work to understand their origin and their role in determining anomalous physical properties of high temperature superconductors.

We acknowledge helpful discussions with J. R. Shi, T. Xiang, Z.Y. Weng, and S. Kivelson. This work is supported by the NSFC, the MOST of China (973 project Nos. 2006CB601002, 2006CB921302), and CAS (Projects IT-SNEM and 100-Talent). The work at BNL is supported by the DOE under Contract No. DE-AC02- 98CH10886.

[1]  A. Damascelli *et al.*, Rev. Mod. Phys. **75**, 473 (2003); J. C. Campuzano *et al.*, in *The Physics of Superconductors*, edited by K. H. Bennemann and J. B. Ketterson (Springer, New York, 2004), Vol. 2; X. J. Zhou *et al.*, in *Handbook of High-Temperature Superconductivity: Theory and Experiment*, edited by J. R. Schrieffer (Springer,NewYork, 2007).

[2]  P. V. Bogdanov *et al.*, Phys. Rev. Lett. **85**, 2581 (2000).

[3]  P. Johnson *et al.*, Phys. Rev. Lett. **87**, 177007 (2001).

[4]  A. Kaminski *et al.*, Phys. Rev. Lett. **86**, 1070 (2001).

[5]  A. Lanzara *et al.*, Nature **412**, 510 (2001).

[6]  X. J. Zhou *et al.*, Nature **423**, 398 (2003).

[7]  A. A. Kordyuk *et al.*, Phys. Rev. Lett. **97**, 017002 (2006).

[8]  A. D. Gromko *et al.*, Phys. Rev. B **68**, 174520 (2003).

[9]  T. K. Kim *et al.*, Phys. Rev. Lett. **91**, 167002 (2003).

[10]  T. Cuk *et al.*, Phys. Rev. Lett. **93**, 117003 (2004).

[11]  M. Eschrig and M. R. Norman, Phys. Rev. Lett. **85**, 3261 (2000).

[12]  F. Ronning *et al.*, Phys. Rev. B **71**, 094518 (2005); J. Graf *et al.*, Phys. Rev. Lett. **98**, 067004 (2007); B. P. Xie *et al.*, Phys. Rev. Lett. **98**, 147001 (2007); T. Valla *et al.*, Phys. Rev. Lett. **98**, 167003 (2007); W. Meevasana *et al.*, Phys. Rev. B **75**, 174506 (2007); J. Chang *et al.*, Phys. Rev. B **75**, 224508 (2007); D. S. Inosov *et al.*, Phys. Rev. Lett. **99**, 237002 (2007).

[13]  G. D. Liu *et al.*, Rev. Sci. Instrum. **79**, 023105 (2008).

[14]  A. A. Kordyuk *et al.*, Phys. Rev. B **71**, 214513 (2005).

[15]  X. J. Zhou *et al.*, Phys. Rev. Lett. **95**, 117001 (2005).

[16]  R. J. McQueeney *et al.*, Phys. Rev. Lett. **87**, 077001 (2001).

[17]  H. He *et al.*, Phys. Rev. Lett. **86**, 1610 (2001).

[18]  M. Hengsberger *et al.*, Phys. Rev. B **60**, 10 796 (1999).

[19]  A. W. Sandvik *et al.*, Phys. Rev. B **69**, 094523 (2004).

[20]  S. Engelsberg and J. R. Schrieffer, Phys. Rev. **131**, 993 (1963).

[21]  H. F. Fong *et al.*, Phys. Rev. B **61**, 14 773 (2000); P. C. Dai *et al.*, Science **284**, 1344 (1999); B. Vignolle *et al.*, Nature Phys. **3**, 163 (2007).

[22]  J. Hwang *et al.*, Phys. Rev. B **75**, 144508 (2007).

# 19

# Specific-Heat Measurement of a Residual Superconducting State in the Normal State of Underdoped $Bi_2Sr_{2-x}La_xCuO_{6+\delta}$ Cuprate Superconductors*

H. H. Wen, G. Mu, H. Q. Luo, H. Yang, L. Shan, C. Ren, P. Cheng, J. Yan and L. Fang

*National Laboratory for Superconductivity, Institute of Physics and Beijing National Laboratory for Condensed Matter Physics, Chinese Academy of Sciences, P.O. Box 603, Beijing 100080, People's Republic of China*

We have measured the magnetic field and temperature dependence of specific heat on $Bi_2Sr_{2-x}La_xCuO_{6+\delta}$ single crystals in wide doping and temperature regions. The superconductivity related specific-heat coefficient $\gamma_{sc}$ and entropy $S_{sc}$ are determined. It is found that $\gamma_{sc}$ has a humplike anomaly at $T_c$ and behaves as a long tail which persists far into the normal state for the underdoped samples, but for the heavily overdoped samples the anomaly ends sharply just near $T_c$. Interestingly, we found that the entropy associated with superconductivity is roughly conserved when and only when the long tail part in the normal state is taken into account for the underdoped samples, indicating the residual superconductivity above $T_c$.

One of the most important issues in cuprate superconductors is the existence of a pseudogap above $T_c$ in the underdoped region [1]. It appears in close relationship with many anomalous properties in the normal state and thus receives heavy debate about its nature. One scenario assumes that the pseu-

---

* Reprinted with permission from H. H. Wen, G. Mu, H. Q. Luo *et al.*, Phys. Rev. Lett. **103**, 067002 (2009). Copyright (2009) by the American Physical Society.

dogap reflects only a competing or coexisting order of superconductivity and it may have nothing to do with the pairing. However, other pictures, such as the Anderson's resonating-valence-bond (RVB) model [2] and related models [3,4], regard the pseudogap as due to the spin-singlet pairing in the spin liquid state, and it has a close relationship with Cooper pairing for superconductivity. Experimentally, some evidence for fluctuating superconductivity in the normal state of underdoped samples has been inferred in the measurements of the Nernst effect [5,6], diamagnetization [7], time-domain optical conductivity [8] and thermal expansion [9], etc. The evidence from specific heat (or entropy) for this residual superconductivity in the normal state is, however, still lacking.

By using the differential heat capacity technique, Loram et al. [10] successfully measured the electronic specific heat (SH) of cuprate superconductors (most of the time at zero field). The advantage of this technique made it possible to observe the SH anomaly near $T_c$ and the suppression to the electronic SH coefficient $\gamma_e$ below $T^*$ in underdoped region. It remains, however, unresolved whether this suppression to $\gamma_e$ below $T^*$ is due to the preformed pairing or induced solely by the competing order [11]. In addition, for a superconductor within the BCS scenario, the superconductivity related entropy (SRE) is conserved at just above $T_c$. We are thus also curious to know whether the SRE is conserved in very underdoped samples. Answering this question casts a big challenge since the SRE is difficult to determine in cuprate superconductors. One way to reach this goal is to measure the difference of heat capacity between the superconducting state and a normal state background which is normally achieved by using a high magnetic field to suppress the superconductivity. The heat capacity under magnetic fields has been measured near $T_c$ by Junod, Erb, and Renner on YBCO, Bi-2212, and Bi-2223 single crystals [12]. Because of the very high critical field in those samples, the relatively low magnetic field (about 10 T) in the usual laboratory cannot suppress the bulk superconductivity completely. It is thus highly desired to do the fielddependent SH measurement on some single crystals with low $T_c$; in such a case, a magnetic field in the scale of 10 T can suppress the bulk superconductivity. As far as we know, no such investigations on SH on systematic doped cuprate samples have been reported. In this Letter, we present the SH data measured on high quality $Bi_2Sr_{2-x}La_xCuO_6$ (Bi-2201) single crystals [13] in a wide temperature and doping regime, and the superconductivity is tuned by the magnetic field. The evidence for residual superconductivity far above $T_c$ has been found based on the analysis of entropy conservation in underdoped samples.

In this experiment, we have selected six high quality crystals grown by the traveling solvent floating zone technique [13]; five of them are from $Bi_2Sr_{2-x}La_x$ $CuO_{6+\delta}$ with $x = 0.8$ (underdoped, $p \approx 0.11$, $T_c = 11$ K), $x = 0.7$ (underdoped, $p \approx 0.123$, $T_c = 18.5$ K), $x = 0.6$ (underdoped, $p \approx 0.131$, $T_c = 22$ K), $x = 0.4$ (optimally doped, $p \approx 0.16$, $T_c = 30$ K), and $x = 0.1$ (overdoped, $p \approx 0.20$, $T_c = 17.6$ K), and one of $Bi_{1.74}Sr_{1.88}Pb_{0.38}CuO_{6+\delta}$ (overdoped, $p \approx 0.22$, $T_c = 9.4$ K). For simplicity, they are denoted as UD11K, UD18.5K, UD22K, OP30K, OD17.6K, and OD9.4K, respectively. In Fig. 19.1, we present the ac susceptibility of two underdoped samples in (a) and (b) and one overdoped sample (with Pb doping) in (c). For the underdoped samples [see, for example, Fig. 19.1(b)], a very small magnetic field can suppress the superconducting transition quickly manifesting a very fragile superfluid density. If we take the point where both the real part susceptibility $\chi'$ and the imaginary part $\chi''$ merge into the flattened normal state background (actually buried in the noise level) as the criterion for bulk superconductivity, the critical field $H^*(T)$ is obtained and shown in Fig. 19.1(d). One can see that, when the field is beyond 9 T, no bulk superconductivity can be detected above 2 K. This allows us to use the data at 9 T as the appropriate background for the state without bulk superconductivity above 2 K [14]. Thus we define the superconductivity related SH as $\gamma_{sc} = [C(H) - C(9\ T)]/T$, where $C(H)$ and $C(9\ T)$ are the total heat capacity measured at the magnetic field $H$ and 9 T, respectively. This treatment naturally removes the phonon contribution since it is field-independent.

Figure. 19.2 presents the temperature dependence of $\gamma_{sc}$ for the corresponding samples shown in Fig. 19.1. The heat capacity was measured by using the relaxation method based on a physical property measurement system (Quantum Design) with the latest upgraded puck. For the underdoped samples, one can easily draw the following interesting conclusions: (1) In the zero temperature approach, the magnetic field always enhances $\gamma_{sc}$, leading to a finite quasiparticle density of states. This is consistent with the results in $La_{2-x}Sr_xCuO_4$ and other systems [15,16]. Our results support also the

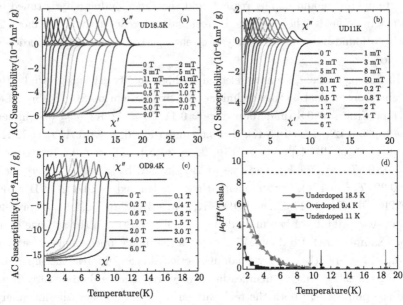

Fig. 19.1   ac susceptibility for three single crystals of (a) UD18.5K, (b) UD11K, and (c) OD9.4K. The measurements were done with an ac field of 0.1 Oe and an oscillating frequency of 333 Hz. The critical field $H^*$ for bulk superconductivity (see text) is shown in (d). The arrows indicate the positions of the bulk superconducting transitions at zero field for the three samples. In this study, all measurements were done with the magnetic field parallel to the $c$ axis of the crystals.

conclusion of a Fermi surface in the normal state revealed by recent quantum oscillation measurements [17]. (2) What surprises us is that there is *no* steplike SH anomaly at $T_c$ for the underdoped samples; instead, it shows a broad humplike peak at about $T_c$ and remains as a long tail of $\gamma_{sc}(T)$ far above $T_c$. For example, for the underdoped sample with $T_c = 11$ K, this long tail can last up to about $42 \pm 5$ K, where the signal is buried in the noise background. (3) In a BCS superconductor, when the superconductivity is suppressed by a magnetic field, the peak height of the SH anomaly is suppressed and the transition temperature is lowered due to the field induced pair breaking [see an example in Fig. 19.2(d) for a conventional BCS superconductor Nb]. However, as shown in Figs. 19.2(a) and 2(b), for the underdoped samples, one can see that the

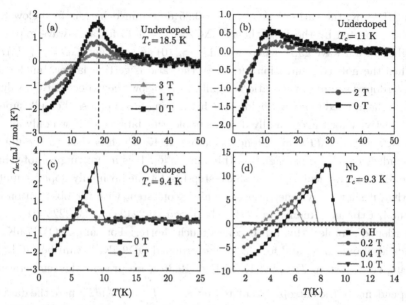

Fig. 19.2   The subtracted specific heat for four samples: (a) UD18.5K, (b) UD11K, (c) OD9.4K, and (d) Nb with $T_c = 9.3$ K(using 2 T as the background). In (a) and (b) the dashed lines mark the positions of $T_c$.

position of the SH peak remains unchanged but the height is suppressed greatly by the magnetic field. Very surprisingly, the onset for bulk superconductivity as measured by the ac susceptibility shifts quickly with the magnetic field. This indicates that the bulk superconductivity is not determined by the position of the SH anomaly. Regarding the long tail of $\gamma_{sc}(T)$ extending up to high temperatures, we conclude that there is residual superconductivity far above $T_c$. In order to check whether this is a special property for the underdoped samples, in Fig. 19.2(c), we present the data for a heavily overdoped sample in the same system. It is easy to see that the $\gamma_{sc}(T)$ data show only a steplike BCS mean field transition with the absence of the long tail in the normal state.

To further illustrate the difference between the underdoped and overdoped samples, we present the $\gamma_{sc}(T)$ data in Figs. 19.3(a) and 19.3(b). For underdoped samples, the long tail of $\gamma_{sc}(T)$ extends to the temperature region between 35 and 45 K. In addition, towards underdoping, the SH peak is strongly suppressed leading to a humplike anomaly. For the strongly underdoped sample UD11 K, the ratio of $\Delta C / \gamma_n T_c = 0.25$ determined here is far below the value expected by the BCS theory ($\Delta C / \gamma_n T_c = 1.43$ for an $s$-wave gap and higher for a $d$-wave gap), where we take $-\gamma_{\mathrm{sc}}(0)$ as $\gamma_n(0)$ and $\Delta C = \gamma_{\mathrm{sc}}(T_c)T_c$. When the hole concentration increases, the ratio is getting larger, but for all underdoped samples, this ratio is significantly below the expected BCS value. Since the applied magnetic field is not high enough to suppress the bulk superconductivity for the optimally doped sample, the data were shown only above 15 K, and the $\gamma_{\mathrm{sc}}(T)$ tail extends to about 42 K, which is close to the upper boundary of the Nernst signal in this sample [6]. It is interesting to note that the SH anomaly near $T_c$ is not sharp-step-like for the optimally doped sample; rather, it shows a symmetric peak. This is consistent with the observation by Junod, Erb, and Renner in optimally doped Bi-2212 and Bi-2223 [18]. For overdoped samples, this tail becomes much shorter: For sample OD17.6K, it ends at about 23 K, and for the very overdoped OD9.4K, it vanishes at 10 K being very close to $T_c = 9.4$ K. In Fig. 19.3(c), we present the temperature dependence of the entropy calculated by $S_{\mathrm{sc}} = \int_0^T \gamma_{\mathrm{sc}}(T')dT'$; here the data of $\gamma_{\mathrm{sc}}(T)$ at $T = 0$ K were obtained by doing the linear extrapolation of the low temperature data. For the overdoped sample OD9.4K, the entropy is conserved at just $T_c = 9.4$ K. The slight nonzero entropy above $T_c$ is induced by the uncertainty in deriving the value of $\gamma_{\mathrm{sc}}(T)$ at $T = 0$ K. The condensation energy calculated by integrating the entropy, i.e., $E_{\mathrm{cond}} = -\int_0^{T_c} S_{\mathrm{sc}}(T')dT'$, is about $48 \pm 5$ mJ/mol for sample OD9.4K. For the underdoped sample UD18.5K, the entropy is obviously not conserved by integrating $\gamma_{\mathrm{sc}}(T)$ just up to $T_c$, but, surprisingly, it becomes roughly conserved when the long tail part of $\gamma_{\mathrm{sc}}(T)$ in the normal state is taken into account as shown by the red circles in Fig. 19.3(c). As stressed previously [19,20], in underdoped cuprates, the term "condensation energy" may have a different meaning as compared to a conventional superconductor since the pairing in the normal state certainly contributes a significant part to the total condensation energy, but the bulk superconducting transition at $T_c$ saves extra energy. By integrating the entropy from $T$ to 50 K, namely, $E_{\mathrm{cond}} = -\int_T^{50K} S_{\mathrm{sc}}(T')dT'$, we derived the temperature dependence of the condensation energy $E_{\mathrm{cond}}$ for three underdoped samples UD11K, UD18.5K, and UD22K and the heavily overdoped sample OD9.4K (integral

from $T$ to 18 K). The results are shown in Fig. 19.3(d). For sample UD18.5K
the total condensation energy at $T = 0$ K is about $263\pm10$ mJ=mol, while the
normal state contributes an energy savings of about $52\pm5$ mJ=mol; this gives
a portion of about 20% of the total condensation energy. An estimate for the
more underdoped sample UD11K finds that the normal state contribution to
the total condensation energy can be as large as 54%, as shown by the blue
triangles in Fig. 19.3(d). This large ratio of the normal state contribution to
the condensation energy makes it almost impossible to attribute the residual
superconductivity above $T_c$ to the Gaussian fluctuation. It also clearly pro-
hibits us from understanding the superconducting transition in underdoped
samples within the BCS scenario.

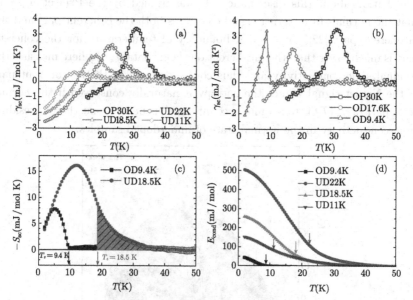

Fig. 19.3   A collection of $\gamma_{sc}(T)$ at zero field for three underdoped samples and one
optimally doped sample in (a) and two overdoped samples and one optimally doped
sample in (b). (c) Temperature dependence of the superconductivity related entropy
calculated by integrating $\gamma_{sc}(T)$ in a wide temperature region. (d) The condensation
energy calculated through integrating the entropy (see text). The arrows mark the
temperatures of the bulk superconducting transition.

In Fig. 19.4, we present a generic phase diagram derived from our data. Here we used the empirical relation $p = 0.21 - 0.18\ x$ to obtain the hole concentration [21]. The red squares represent the $T_c$ values of our samples, which show very good consistency with that of Ando *et al.* [21]. The blue circles show the vanishing points $T_{SH}$ of the long tail of $\gamma_{sc}(T)$ using the criterion of $0\pm0.15$ mJ/molK$^2$, where the SRE has dropped below 0.5 mJ/molK [see Fig. 19.3(c)]. One can see that the difference between $T_c$ and $T_{SH}$ is getting monotonically larger towards underdoping. This phase diagram looks qualitatively similar to that depicted based on the Nernst measurements [6,22], but the upper limit temperatures for the Nernst signal on underdoped samples are higher than the values derived from our specific heat. There is a possible explanation about this discrepancy: It was argued by the Princeton group that the normal state Nernst signal comprises both the coherent part and the incoherent part [22]. The upper boundary of temperature for the coherent part is much lower than the incoherent one. Our data $\gamma_{sc}(T)$ here measure the residual superconductivity and thus correspond well with the coherent part of the Nernst signal. Since the entropy is naturally conserved if the normal state part of $\gamma_{sc}(T)$ is taken into account, we thus believe that there is residual superconductivity in the normal state of underdoped samples.

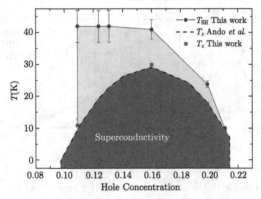

Fig. 19.4  A generic phase diagram plotted based on the specific-heat data. The dashed line is the $T_c$-$p$ plot from Ando's group in the same system. The red squares represent the measured $T_c$ values of our samples at the same nominal doping level. The blue circles show the temperatures $T_{SH}$ where $\gamma_{sc}(T) = 0 \pm 0.15$ mJ/molK$^2$ (within the error bars of the experiment). One can see that the gap between $T_c$ and $T_{SH}$ is getting monotonically larger but $T_{SH}$ flattens out in more underdoped region.

Our results are also qualitatively consistent with the recent observation of local pairing above $T_c$ as seen by scanning tunneling microscopy [23]. These nanoscale droplets of Cooper pairs above $T_c$ will certainly contribute to the condensation energy of the system and make the entropy unconserved (at $T_c$) unless the upper temperature for counting the entropy is beyond $T_{SH}$ in our definition. In this sense the superconducting transition in underdoped samples means to establish the long range phase coherence [3]. Thus the thermal energy $k_B T_c$ may be equated by the phase coherence energy $E_{coh} = \hbar^2 \rho_s(T_c)/m^*$ given by Deutscher [24], where $\rho_s$ is the superfluid density and $m^*$ is the effective mass. Below $T_c$, the quasiparticles which reside on the small Fermi surfaces in the normal state [17,25] will pair up with each other and condense into the superconducting state together with the residual Cooper pairs formed above $T_c$. This naturally builds up a new gap on the small Fermi surfaces in the region near the nodes [26,27]. Above $T_c$, strong phase fluctuation [3,28] breaks up many Cooper pairs with small pairing energy [25], but some residual pairs with stronger pairing strength will exist up to a high temperature. As demonstrated by our data, the superconducting condensation in the underdoped region cannot be put into the BCS category.

In summary, the specific-heat anomaly at $T_c$ is strongly suppressed through underdoping leading to a humplike anomaly with the height much below the value predicted by the BCS theory. A long tail of $\gamma_{sc}(T)$ has been found far into the normal state for underdoped samples. The entropy calculated by integrating $\gamma_{sc}(T)$ to $T_c$ is clearly not conserved, but it becomes roughly conserved when and only when the tail part in the normal state is taken into account. These results prohibit from using the BCS picture to understand the superconducting transitions in underdoped samples.

We thank J. Zaanen, J. Tallon, and P. W. Anderson for comments and suggestions. We acknowledge also S. Kivelson, F. C. Zhang, Z. Y. Weng, Q. H. Wang, G. Appeli, P. C. Dai, and Y. Y. Wang for useful discussions. We thank L. Zhao, G. D. Liu, and X. J. Zhou for providing us one as-grown sample (OD9.4K). This work was supported by the MOST of China (973 Projects No. 2006CB601000 and No. 2006CB921802) and CAS Project.

[1]   T. Timusk and B.W. Statt, Rep. Prog. Phys. **62**, 61 (1999).

[2]   For a review on the RVB picture, see P. W. Anderson, P. A. Lee, M. Randeria, T. M. Rice, N. Trivedi and F. C. Zhang, J. Phys. Condens. Matter **16**, R755 (2004).

[3]   S. A. Kivelson and V. J. Emery, Nature **374**, 434 (1995).

[4]   Z. Y. Weng, D. N. Sheng, and C. S. Ting, Phys. Rev. Lett. **80**, 5401 (1998).

[5]   Z. A. Xu *et al.*, Nature **406**, 486 (2000).

[6]   Y. Wang, L. Li and N. P. Ong, Phys. Rev. B **73**, 024510 (2006).

[7]   Y. Wang *et al.*, Phys. Rev. Lett. **95**, 247002 (2005).

[8]   J. Corson *et al.*, Nature **398**, 221 (1999); A. J. Millis, Nature **398**, 193 (1999).

[9]   C. Meingast *et al.*, Phys. Rev. Lett. **86**, 1606 (2001).

[10]  J. W. Loram *et al.*, Phys. Rev. Lett. **71**, 1740 (1993); J. Phys. Chem. Solids **62**, 59 (2001).

[11]  J. L. Tallon *et al.*, J. Phys. Chem. Solids **59**, 2145 (1998).

[12]  A. Junod, A. Erb and C. Renner, Physica **317C**, 333 (1999).

[13]  H. Q. Luo *et al.*, Supercond. Sci. Technol. **21**, 125024 (2008).

[14]  To define a "normal state background" is a nontrivial issue for underdoped cuprate superconductors. Here we use the data measured at 9 T as a relative but appropriate background because the ac susceptibility shows no trace of bulk superconductivity above 2 K. Furthermore, we found that the difference between the specific-heat data of 9 T and 8 T is almost invisible.

[15]  H. H. Wen *et al.*, Phys. Rev. B **72**, 134507 (2005).

[16]  H. H.Wen and X. G.Wen, Physica **460C**, 28 (2007).

[17]  N. Doiron-Leyraud *et al.*, Nature **447**, 565 (2007).

[18]  J. W. Loram, J. L. Tallon, and W.Y. Liang, Phys. Rev. B **69**, 060502(R) (2004).

[19]  S. Chakravarty, Phys. Rev. Lett. **82**, 2366 (1999); S. Chakravarty and H.Y. Kee, Phys. Rev. B **61**, 14821 (2000).

[20]  D. Van der Marel *et al.*, Phys. Rev. B **66**, 140501 (2002).

[21]  Y. Ando *et al.*, Phys. Rev. B **61**, R14956 (2000).

[22]  Y. Wang, Ph.D. dissertation, Princeton University, 2004.

[23]  K. K. Gomes *et al.*, Nature **447**, 569 (2007).

[24]  G. Deutscher, Nature **397**, 410 (1999).

[25]  A. Kanigel *et al.*, Phys. Rev. Lett. **99**, 157001 (2007).

[26]  H. Matsui *et al.*, Phys. Rev. Lett. **90**, 217002 (2003).

[27]  W. S. Lee *et al.*, Nature **450**, 81 (2007).

[28]  Z. Tesanovic, Phys. Rev. B **59**, 6449 (1999).

# 20

# A Brief Overview on Raman Scattering in Cuprate Superconductors*

M. J. Wang, A. M. Zhang and Q. M. Zhang

*Department of Physics, Renmin University, Beijing 100872, P. R. China*

High-temperature superconductivity is one of the most fascinating and challenging areas of condensed matter physics. In the past two decades, a huge number of theoretical and experimental studies have been contributed to the field and many crucial progresses have been made. Raman scattering has been extensively applied to high-$T_c$ superconductors owing to its unique abilities to probe multiple primary excitations and symmetry selection rules. In this chapter, we present a brief review on Raman scattering experimental achievements in cuprate superconductors, with emphasis on pairing mechanism related aspects such as electron-phonon coupling, pair-breaking peak and two-magnon etc. And also we will address some open questions.

## 20.1 Introduction

When light propagates through a medium, it will interact with microscopic particles or elementary excitations in medium, leading to a change of energy and momentum—this phenomenon is called light scattering. If scattered light has an identical frequency with incident light, it is an elastic scattering, or the so-called Rayleigh scattering, otherwise it is called inelastic light scattering. There are a lot of excitations which can cause inelastic scattering in solids, such as optical phonons, acoustic phonons, electrons, magnons and many other kinds of elementary excitations. Generally, we call such inelastic scattering as Raman scattering. For the scattering with a very small energy transition

---

* Based on the talk presented at the Symposium on High $T_c$ Superconductivity, 2008 CCAST.

($\sim$2-3 cm$^{-1}$), it is called Brillouin scattering. Raman scattering is different from infrared absorption, where a monochromatic light as incident light is required. After the incident light enters a solid, it will interact with elementary excitations and an energy transfer will happen. The Raman shift we usually see in a Raman spectrum, is exactly the transfer energy between incident light and elementary excitations. Raman scattering can be classified into Stokes and anti-Stokes procedures, corresponding to energy transfer from incident light to elementary excitation and vice versa.

Raman scattering has been widely applied in fundamental research fields such as physics, chemistry, biography etc., and many industrial areas. Basically it can be used to characterize and measure rotating and vibrating modes, by which one can study composition, phase transition, stress etc., in condensed matter. In addition, Raman scattering also can be used to study the properties related to the electrons, which originates from the scattering by electrons or holes near the Fermi surface. Raman scattering is also a good technique to detect magnetic excitation. Electronic Raman scattering (ERS) had very limited applications in the 1960s and 1970s because the penetration depth is very small in normal metals so the effective intensities of incident light are low, and Coulomb screening effect is very strong due to high density of electrons in metals. Electronic Raman scattering reflects charge density-density correlation, but the strong screening in metals will much weaken density correlation effect. High temperature superconductors (HTSC) are suitable for electronic Raman scattering due to its lower density of charge carriers. Moreover, the Cu-O planes in HTSC have a basic tetragonal symmetry, which is a great advantage that can be taken by Raman scattering. One can study the anisotropic excitations from nodes and antinodes of Brillouin zone (BZ) by symmetry analysis, which is particularly useful in exploring $d$-wave pairing in cuprate superconductors. Today, ERS has become one of the most important experimental techniques in studying HTSC [1,2].

## 20.2  Electron Raman Scattering in High Temperature Superconductors

### 20.2.1  Phenomenological and Microscopic Theoretical Pictures

According to the classical theory of electromagnetic field, atoms (molecules) will be polarized by light radiation field entering into solid, inducing vibrating dipoles and instantaneous polarization. Instantaneous polarization can be described as a linear plus a nonlinear term as following,

$$P = \varepsilon_0(\chi E_I + \chi' X E_I).$$

The first term corresponds to elastic scattering (Rayleigh scattering) and the second term represents inelastic scattering that changes the frequency of incident light.

Phenomenologically the intensity of inelastic scattering can be described by linear response theory. The response from a linear passive system can be decomposed to a number of damped harmonic oscillators. We define the generalized displacement as

$$\overline{X}(q,\omega) = T(q,\omega)F(\omega).$$

Using fluctuation-dissipation theorem,

$$< X(q)X^+(q) >_\omega = \frac{\hbar}{\pi}\{n(\omega) + 1\}\mathrm{Im}T(q,\omega),$$

we can see that Raman scattering is a measure of the imaginary part of response function. For ERS, the effective Raman response function reflects the response of charge system to the perturbation by light radiation field. Electronic Raman response function is dominated by vertex function between light radiation field and electrons, which is determined by band structure and crystal symmetry. The theoretical details of ERS can be found in [1,2].

One of the common features for HTSC is the same Cu-O squares. In general, it is treated as $D_{4h}$ point group though in some cases there may exist a small deviation from the ideal squares. The following discussions are based on this symmetry.

According to general phenomenological theory, the intensity of Raman scattering is proportional to the square of the matrix of second-order polarizability tensors. The Raman-active irreducible representations of $D_{4h}$ point group are listed here, as well as the corresponding polarizability tensors, $x$, $y$

and $z$ representing the principal axes determined by second-order polarizability tensors.

So the modes with different symmetries are selected by polarization geometries. Usually we use symbols like $z(x, y)z$ to represent the experimental geometry. The first and last letters give the propagation directions of incident and scattered light, respectively. And the two letters in the bracket indicate the polarizations of incident and scattered light, respectively. We can also detect the combination of different modes by selecting polarizations of incident and scattered light if it is difficult to directly obtain a pure symmetry by polarization geometry. The polarization geometries for HTSC are shown in Fig. 20.1, in which polarizations of incident and scattered light are indicated by the arrows. The third ($A_{2g} + B_{1g}$) and fourth ($A_{2g} + B_{2g}$) cases are two kinds of vertical geometries which are used most frequently. Generally Raman intensity is dominated by the contribution from $B_{1g}$ channel in the third

$$
\begin{array}{ccccc}
A_{1g} & A_{2g} & B_{1g} & B_{2g} & E_g \\
\begin{bmatrix} a & & \\ & a & \\ & & b \end{bmatrix} &
\begin{bmatrix} & c & \\ -c & & \\ & & \end{bmatrix} &
\begin{bmatrix} d & & \\ & -d & \\ & & \end{bmatrix} &
\begin{bmatrix} & e & \\ e & & \\ & & \end{bmatrix} &
\begin{bmatrix} & & f \\ & & \\ g & & \end{bmatrix}
\end{array}
$$

$$
x = \begin{pmatrix} 1 \\ 0 \\ 0 \end{pmatrix} \quad
y = \begin{pmatrix} 0 \\ 1 \\ 0 \end{pmatrix} \quad
z = \begin{pmatrix} 0 \\ 0 \\ 1 \end{pmatrix}
$$

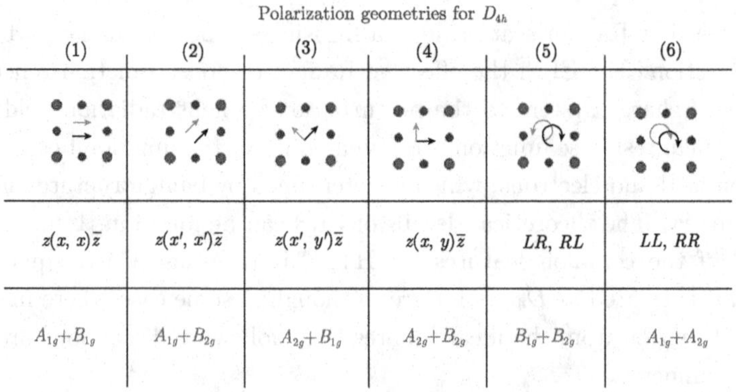

Polarization geometries for $D_{4h}$

| (1) | (2) | (3) | (4) | (5) | (6) |
|-----|-----|-----|-----|-----|-----|
| | | | | | |
| $z(x, x)\bar{z}$ | $z(x', x')\bar{z}$ | $z(x', y')\bar{z}$ | $z(x, y)\bar{z}$ | LR, RL | LL, RR |
| $A_{1g}+B_{1g}$ | $A_{1g}+B_{2g}$ | $A_{2g}+B_{1g}$ | $A_{2g}+B_{2g}$ | $B_{1g}+B_{2g}$ | $A_{1g}+A_{2g}$ |

Fig. 20.1  Raman tensors and six polarization geometries for $D_{4h}$ point group in cuprate superconductors. In the lower panel, red spots represent Cu atoms and blue ones represent O atoms. The first four are linear polarization geometries, the last two are circular polarization ones. The third and fourth cases are particularly important because approximately they can be considered as $B_{1g}$ and $B_{2g}$ contributions.

geometry because the contribution of $A_{2g}$ is negligible below 1000 cm$^{-1}$. So the third geometry can be considered as $B_{1g}$ symmetry. Similarly the fourth case in Fig. 20.1 is called as $B_{2g}$ symmetry. Circular polarization geometries are illustrated in the fifth and sixth cases in Fig. 20.1, which is particularly helpful when exploring some circular excitations such as phonic or magnetic chiral modes. Using all of the six geometries one can makes a complete symmetry analysis to precisely separate a pure spectrum for each symmetry.

The above example of cuprate superconductors introduces formal symmetry phenomenologically. In a microscopic view, the cross section of ERS is closely related to its structural factor, formally which is similar to neutron scattering. In principle the cross section is proportional to the imaginary part of correlation function. When a solid is irradiated by an electromagnetic wave, oscillatory electric field $E$ will cause a perturbation for charge system, leading to an effective charge-density fluctuation, which reflects an interaction between charge and light radiation field. Effective fluctuation of charge system is determined by interaction vertex with light, which is a key of electronic Raman scattering. Theoretically, in the absence of inter band transition the vertex is proportional to the reciprocal of mass tensor, i.e., second derivatives of band dispersion. Considering HTSC as a tetragonal symmetry, we can expand the vertex in an analytical way. Taking the first-order term, the vertex for $B_{1g}$ can be written as

$$\gamma_k = \cos(k_x) - \cos(k_y)$$

and the vertex of $B_{2g}$ can be written as

$$\gamma_k = \sin(k_x)\sin(k_y),$$

which have the same symmetry as $d$-wave. Fig. 20.2 shows the relative amplitudes of the two vertex in the first BZ. $B_{1g}$ and $B_{2g}$ channels pick up quasiparticles excitation at antinodes $(\pm\pi, 0)$ and nodes $(\pm\pi/2, \pm\pi/2)$, respectively. This is a unique advantage for ERS in the study of HTSC, corresponding to $45°$ angle-resolution anisotropy. Note that the spectra from two channels correspond to low-energy excitations not exactly at the nodes and antinodes in $k$ space, but a convolution in the region around the nodes and antinodes.

In normal state electronic Raman scattering can be seen as a Drude process similar to infrared conductivity, whereas a pair-breaking peak is a prominent feature in superconducting state which is contributed by breaking cooper pairs

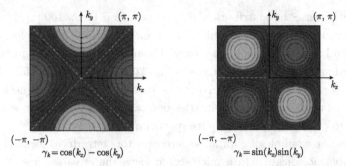

Fig. 20.2   The amplitudes of the vertex of $B_{1g}$ (left) and $B_{2g}$ (right) in the first BZ, which indicates $B_{1g}$ and $B_{2g}$ electronic Raman spectra detect antinodal and nodal regions, respectively.

with incident light. For an isotropic conventional BCS superconductor, a uniform pair-breaking peak is expected exactly at $2\Delta_0$, while it is quite different for a $d$-wave superconductor. A well-established ERS theory for $d$-wave superconductors predicts that a $d$-wave superconductor will show a pair-breaking peak at $2\Delta_0$ in $B_{1g}$ channel and below $2\Delta_0$ ($\sim 1.7\Delta_0$) in $B_{2g}$ channel. The $d$-wave anisotropy is exactly resolved by ERS, as we mentioned above. Furthermore, low-energy behavior is power-law form for $B_{1g}$ and linear for $B_{2g}$ in the limit of $\Omega \to 0$, while it is an exponential decay at low energies for a conventional $s$-wave superconductor (see Fig. 20.3).

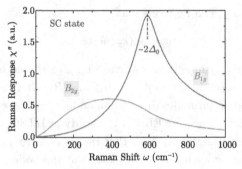

Fig. 20.3   Electronic Raman spectra below $T_c$ in $B_{1g}$ and $B_{2g}$ channels based on $d$-wave superconducting theory. $B_{1g}$ and $B_{2g}$ pair-breaking peaks show an apparent anisotropy, which appear at $2\Delta_0$ and below $2\Delta_0$, and with linear and power-law decays at low energies, respectively.

## 20.2.2    Superconducting State

### 20.2.2.1    Superconducting Gap and Anisotropic Pairing

A good-quality surface can be obtained for Bi-2212 system because it is a layered structure and easy to be cleaved. Many Raman scattering measurements were performed on this system and some consistent results have been reached. Fig. 20.4 shows typical Raman scattering results of Bi-2212 system from two groups. The sharp peaks are attributed to phonons [4], and the results from two groups are basically the same after subtracting phonon peaks.

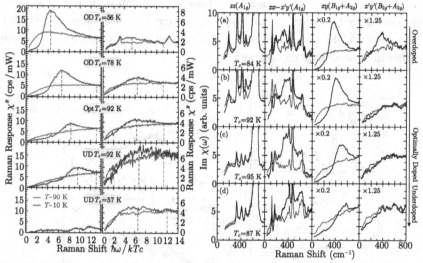

Fig. 20.4    Raman scattering results in Bi2212 system with different doping levels from two groups. Red and blue colors denote normal and superconducting state. The thin line represents normal state and the thick one superconducting state. The left panel is reprinted with permission from [3]. Copyright (2002) by the American Physical Society. The right panel is reprinted with permission from [4], copyright (1999) by the American Physical Society.

In the left panel of Fig. 20.4, from up to down doping level changes from overdoping to underdoping. For electronic Raman spectra of $B_{1g}$ symmetry, there is a clear pair-breaking peak in overdoped region. The shape of pair-breaking peak becomes much broader and its position shifts to a higher frequency with the evolution from overdoping to underdoping[3]. The theory of electronic Raman scattering for cuprate superconductors, points out that $B_{1g}$ peak position corresponds to twice of maximum superconducting energy gap. And the ratio of superconducting gap to $k_B T_c$ continuously increases from 4 to 9 when doping level evolves from overdoping to underdoping, showing an essential difference from conventional BCS superconductors. The significant deviation from 3.52 given by BCS in underdoped region, implies an unconventional superconducting mechanism. $B_{2g}$ peak position is hard to be determined precisely because the peak is much broader. Some reports suggested that $B_{2g}$ peak position traces doping dependence of $T_c$. We will come back to this later when discussing the topic on two-gap structure. The obvious anisotropy between $B_{1g}$ and $B_{2g}$ is consistent with other experiments like angle-resolved photoemission, which strongly supports a scenario of $d$-wave pairing in cuprate superconductors.

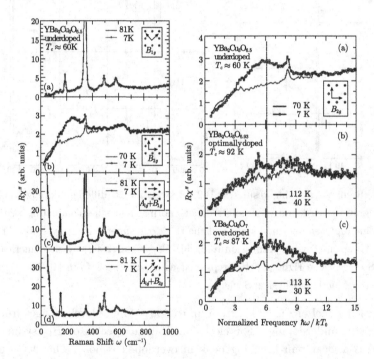

Fig. 20.5   ERS of YBCO system. Reprinted with permission from [5]. Copyright (2000) by the American Physical Society.

The results from YBaCuO system are shown in Fig. 20.5. It is not easy to separate electronic contributions from the $B_{1g}$ spectra because there exists a strong $B_{1g}$ phonon around 340 cm$^{-1}$ (so-called buckling mode). $B_{2g}$ pair-breaking peak still can be observed, which is very broad, similar to that of Bi2212 system [5].

Fig. 20.6 shows ERS from LaSrCuO, which is different from other copper oxide system. It looks like there occurs an exchange in intensity between $B_{1g}$ and $B_{2g}$ spectra, i.e., $B_{2g}$ pair-breaking peak is much stronger than that in $B_{1g}$ channel [6,7]. This is related to electronic stripe phase which we will discuss later.

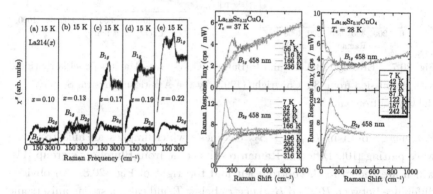

Fig. 20.6    ERS of LSCO system. The left panel is reprinted with permission from [6]. Copyright (1999) by the American physical society. The right panel is reprinted with permission from [7]. Copyright (2002) by the American Physical Society.

In Fig. 20.7 ERS of TlBaCuO and HgBaCaCuO is presented. We can see that pair-breaking peak in $B_{1g}$ channel is clear and the magnitude of peak position over $T_c$ is the same as Bi2212, while we can barely see pair-breaking peak in $B_{2g}$ channel [8]. The clear peak in $A_{1g}$ channel below $T_c$ is considered to be related to spin resonance mode observed by neutron scattering [7].

Fig. 20.8 is ERS of electron-doped cuprate superconductor NdCeCuO. Controversial conclusions on pairing symmetry are reported based on the existing ERS results. In the top left of Fig. 20.8, an earlier Raman result is shown. One can see that pair-breaking peak gets a rapid asymptotic form in

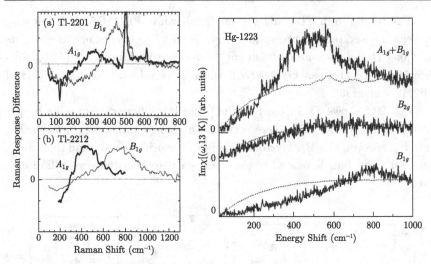

Fig. 20.7    ERS of TlBaCuO and HgBaCaCuO systems. The left panel is reprinted with permission from [8]. Copyright (1997) by the American Physical Society. The right panel from [9].

low energy both in $B_{1g}$ and $B_{2g}$ channels, which implies an anisotropic $s$-wave pairing [10]. Different Raman results come from Blumberg's group [11], which are shown in the bottom left and top right of Fig. 20.8. An obvious difference between $B_{1g}$ and $B_{2g}$ spectra below $T_c$ indicates a strong anisotropic pairing. Interestingly, the $B_{2g}$ peak position is higher than the $B_{1g}$ one, which is opposite to the Bi2212 system, other hole-doped systems and theoretical predictions for a pure $d$-wave. This behavior is explained as an indication of nonmonotonic $d$-wave gap. In the picture of nonmonotonic $d$-wave gap, the maximum of $d$-wave gap appears between the nodal and antinodal regions, as shown in the bottom right panel of Fig. 20.8. This is different from the case of hole-doping where the maximum and zero are located exactly in the antinodal and nodal directions. The shift of the gap maximum in electron-doped systems, gives rise to the different $B_{1g}$ and $B_{2g}$ spectra from hole-doped systems. In a microscopic view, electronic Raman spectra are determined by Raman vertex, hence depend on band structures [12]. Actually a two-band model has been proposed to present a good interpretation on the behavior of ERS in electron-doped cuprate superconductors [1].

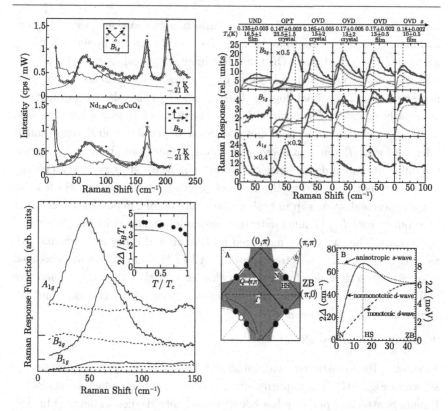

Fig. 20.8    ERS of electron-doped cuprate superconductor NdCeCuO. The upper left panel is reprinted with permission from [10]. Copyright (1995) by the American Physical Society. The upper right panel is reprinted with permission from [11]. Copyright (2005) by the American Physical Society. In which blue color denotes superconducting state. The lower two panels are reprinted with permission from [12]. Copyright (2002) by the American Physical Society. Solid lines represent superconducting state.

## 20.2.2.2    Two Gaps

Recently two-gap structure has attracted much attention in the study of cuprate superconductors. The two energy gaps were first proposed in a paper on electronic Raman scattering in 2006, in which Hg2201 crystals with different doping levels were studied [13]. The paper presents the evolution of pair-breaking peak with varying doping. A well-defined $B_{1g}$ peak below

$T_c$ can be seen in overdoped region, while it gets more and more obscure and monotonically moves to higher energies with decreasing doping level from overdoped to underdoped. On the other hand, $B_{2g}$ peak is quite broad and shows a nonmonotonic dependence on doping level, which reaches a maximum in frequency near optimal doping. In $B_{1g}$ and $B_{2g}$ channels two-gap features are clearly revealed. $B_{2g}$ position follows the dome of $T_c$ vs doping very well, while $B_{1g}$ position shows a monotonic linear dependence on doping, and it intersects with $T_c$ dome in overdoped region rather than optimal doping. So the different doping dependence of $B_{1g}$ and $B_{2g}$ peaks gives rise to two energy scales: one is the gap from $B_{2g}$ channel covering nodal regions, which looks like a superconducting gap in both gap size and doping dependence; the other one comes from $B_{1g}$ channel detecting antinodal regions, which behaves as a pseudogap. The gap values measured by Raman and other experiments such as ARPES, tunneling spectroscopy etc. All of the gap values are classified into two categories: one follows $B_{1g}$ behavior and the other $B_{2g}$. There are still some controversies about the relationship between the two gaps and the superconducting mechanism.

### 20.2.2.3  Impurity Effect

Electronic Raman spectra with magnetic and non-magnetic impurities are shown in Fig. 20.9. The impurity effect in cuprate superconductors on electric Raman Scattering spectrum has been theoretically studied in detail [14]. The main prediction is that low-energy decay in $B_{1g}$ channel below $T_c$ will change from power law to linear by impurity, while the $B_{2g}$ response keeps its linear form at low energies.

Recently there are some new achievements shown in Fig. 20.9. The effect of magnetic impurity is totally different from that of non-magnetic impurity. At low concentrations magnetic impurity has little influence on pair-breaking peaks while the peaks will be completely eliminated by non-magnetic impurity. The $A_{1g} + B_{1g}$ ERS is shown in the left panel of Fig. 20.9, indicating no matter what impurity is doped, the pair-breaking peaks has little change. For pure $B_{1g}$ symmetry, whose ERS is shown in the right panel of Fig. 20.9, the pair-breaking peak is strongly suppressed by Zn doping but survives with Ni-doping. These results lead to a theoretical conjecture based on $t$-$J$ model that $A_{1g}$ electronic Raman spectrum directly reflects superconducting order parameter and $B_{1g}$ represents $d$-wave CDW order parameter [15].

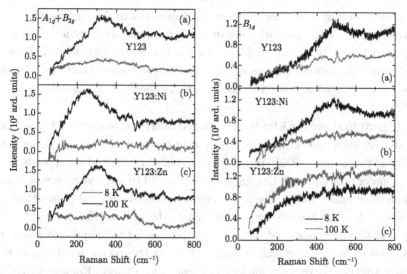

Fig. 20.9    ERS of YBCO system with magnetic and non-magnetic impurities. The left panel is for $A_{1g} + B_{2g}$ channel and the right $B_{1g}$ channel. There is no obvious influence on the pair-breaking peak in $A_{1g} + B_{2g}$ channel by doping Zn and Ni, but Zn doping strongly suppresses pair-breaking peak in $B_{1g}$ channel (right). Reprinted with permission from [14]. Copyright (2004) by the American Physical Society.

Actually there are some different opinions on what contributes to $A_{1g}$ channel. It was argued that there is a correlation between $A_{1g}$ rather than $B_{1g}$ pair-breaking peak and the resonant mode measured by inelastic neutron scattering. This demonstrates that AF spin fluctuations are involved in $A_{1g}$ Raman scattering process in some way or even play a key role.

### 20.2.3    Normal State

#### 20.2.3.1    Pseudogap

The pseudogap problem has been discussed a lot in copper oxide superconductors in both experimental and theoretical aspects. In last two decades, more and more experimental evidences were accumulated and a consensus was reached that there definitely exists a pseudogap in the normal state. However there are still many debates on its origin, not only theoretically but also experimentally. Magnetic measurements such as nuclear magnetic resonance,

neutron scattering and other experiments demonstrated that the pseudogap in
normal state has a spin-relative origin, while transport measurements, tunneling spectrum, ARPES etc., displayed that it is a charge gap. Electric Raman
scattering can measure charge density-density correlation function. In principle, if pseudogap is related to charges, there should be some indications in
ERS.

Fig. 20.10 shows the ERS in underdoped Bi2212 system [16]. A broad peak
located at about 600 cm$^{-1}$ was observed in the normal state. It is not contributed by phonons. It disappears in both normal state and superconducting
state under overdoping. The right panel of Fig. 20.10 displays its temperature
dependence more clearly. The peak is enhanced gradually and passes through
$T_c$ smoothly. The features are considered as the evidence for a normal-state
gap and they are consistent with other experimental techniques. We need to
note that so far there are only a small amount of reports on pseudogap in Raman scattering. It remains unclear how pseudogap affects Raman scattering
in cuprates.

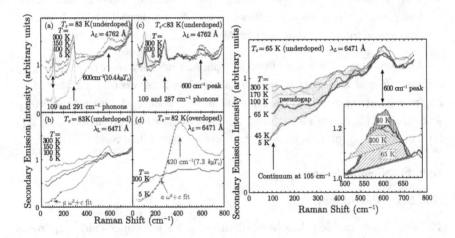

Fig. 20.10    ERS in underdoped Bi2212 system. The left panel is the ERS with
different doping levels. The right panel shows spectra from underdoped samples. A
peak can be seen around 600 cm$^{-1}$ even above $T_c$, which is regarded as a pseudogap
feature. Reprinted with permission from [16].

### 20.2.3.2    Quantum Critical Point

Fig. 20.11 displays the electronic Raman spectra in normal state in $B_{1g}$ and $B_{2g}$ channels in Bi2212 system [17]. As a response function of charge density-density correlation, ERS can present dynamic information on charge

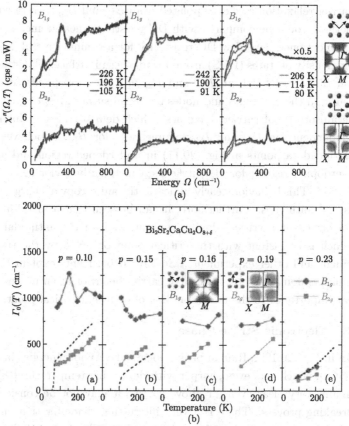

Fig. 20.11    (a) shows ERS of Bi2212 system in normal state at different temperatures. (b) shows the evolution of extrapolated static relaxation rate with doping. Using memory function approach [5], dynamical charge carrier relaxation rate can be extracted from the normal-state ERS and the static relaxation rate can be obtained by extrapolating to zero frequency. The data are compared with transport data (dashed line). Reprinted with permission form [17]. Copyright (2005) by the American Physical Society.

carriers in principle. Using a so-called memory-function approach [5], one can extract a dynamic charge carrier relaxation rate hidden in normal-state Raman spectra. Then static charge carrier relaxation rate can be obtained by extrapolating dynamic charge carrier relaxation rate to zero frequency, as is shown by the red and green date points in Fig. 20.11. In principle, the static charge carrier relaxation rate extrapolated from $B_{1g}$ and $B_{2g}$ spectra with the above method, could be compared with DC resistivity measurements. But it should be mentioned here that DC resistivity for a common metal is just an average in density of states (DOS) around whole Fermi surface, while the static relaxation rate extracted from $B_{1g}$ and $B_{2g}$ spectra corresponds to a DOS integral around the antinodes and nodes on Fermi surface, respectively. So it gives the anisotropy of carrier dynamics, which demonstrates an interesting doping dependence in $B_{1g}$ and $B_{2g}$ symmetries: for $B_{1g}$, it behaves as an insulator (red data points in Fig. 20.11) in underdoped region and a metal in the overdoping region; for $B_{2g}$, it keeps a metallic behavior in the whole doping region. This behavior clearly indicates the anisotropy of charge carriers at nodes and antinodes. Furthermore, it can be seen that a metal-insulator transition occurs at a critical doping concentration in the antinodal region ($B_{1g}$), which is consistent with the critical point observed by the transport measurement and other experiments. So ERS experiments explain that the critical point is dominated by the charge carrier behavior at antinodes, which is of great importance to the understanding of quantum critical point.

### 20.2.3.3 Electronic Stripe Phase

As shown in Fig. 20.12, a Raman peak appears at very low energies in under-doped LSCO, which gets even sharper with decreasing temperature[18]. This peak remains very clear even far above $T_c$. So it may not be considered as a pair-breaking process. There is a very interesting character that the peak appears in $B_{2g}$ channel in heavily underdoped samples ($x = 0.02$) while it turns into $B_{1g}$ channel in lightly underdoped region ($x = 0.10$) near optimal doping.

Combined with neutron scattering and other measurements, this low-energy peak can be understood in term of electronic stripe phases. The sketches are shown in the lower panel of Fig. 20.12, in which the peak is connected with a collective excitation mode of electronic stripes driven by electromagnetic waves. In a heavy underdoping, neutron scattering, $\mu$SR and some other experiments observed electronic stripes in the lower left sketch of Fig. 20.12, whose direction is diagonal, 45° relative to Cu-O bond direction. For $B_{1g}$ symmetry, the polarization of incident light is either parallel or perpendicular to the direction of stripes, so the polarization of radiated scattering light is simply prohibited. Thus one cannot observe the collective excitation mode of electronic stripes. On the other hand, for $B_{2g}$ symmetry the polarization has a component along both incident and scattering directions. In this case the collective excitation of the electronic stripes is permitted. In the lightly underdoped samples, some experiments reveal that the direction of electronic stripes is rotated by 45°, along Cu-O bond direction. Similarly, the collective excitation of electronic stripe is allowed in $B_{1g}$ channel.

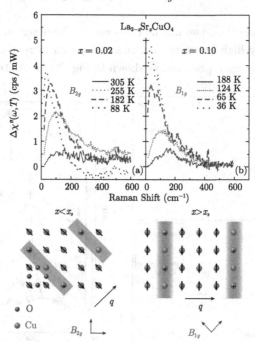

Fig. 20.12    ERS of underdoped LSCO system. One can see low-energy peaks in two underdoped samples. The peak appears in $B_{2g}$ channel in heavily underdoped samples ($x = 0.02$) turns into $B_{1g}$ channel in lightly underdoped region ($x = 0.10$). Reprinted with permission from [18]. Copyright (2005) by the American Physical Society.

## 20.3   Raman Scattering of Two-Magnon Process

In a typical two-dimensional square antiferromagnetic lattice like $K_2NiF_4$, in principle a magnon (or single-spin flip) can be excited after it absorbs incident photons with appropriate energy. However a classical antiferromagnetic spin wave is acoustic-like, which means magnon energy also approaches to zero when its momentum goes down to zero. For this reason a single magnon can hardly be observed in light scattering. However, if two neighboring spins in a $K_2NiF_4$ spin lattice reverse simultaneously in a simple picture, and meet the requirements of energy and symmetry, so-called two-magnon process is permitted in Raman scattering. In a microscopic picture, it looks like a double-exchange procedure between two spins on Cu ions bridged by intermediate oxygen ion.

In cuprate materials, two-magnon process can be observed in $B_{1g}$ channel, which has a very high intensity in antiferromagnetic parent compound, even much higher than most phonons, as shown in Fig. 20.13. As the simultaneous

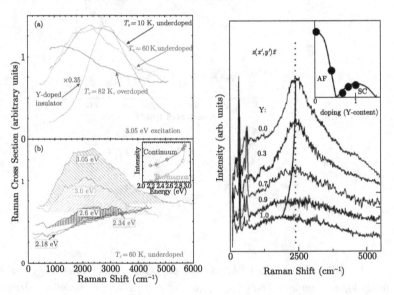

Fig. 20.13   The left panel shows AF two-magnon peaks in Bi2212 samples, and the right in YBCO systems. Both of them show an evolution of two-magnon peak with doping.

reverse of adjacent two spins breaks the nearest six antiferromagnetic couplings, two-magnon peak will appears at about $3J$ in energy. The ratio may slightly deviate from 3, which depends on a detailed microscopic theoretical model. One can see this provides an opportunity to determine AF superexchange energy $J$. Actually it is the most accurate method to measure $J$ in cuprates so far. For insulating parent compounds of cuprates, two-magnon peak is around 3000 cm$^{-1}$, corresponding to a superexchange energy of $\sim$ 1000 cm$^{-1}$ (about 125 meV). With increasing doping level, two-magnon peak becomes broader and moves to lower energies, which denotes the evolution of AF fluctuations with doping [16,19].

A quite low carrier doping in parent compounds will completely destroy long-range AF order. The remaining strong AF spin fluctuations will be further suppressed with approaching optimal doping. Consequently two-magnon signal becomes much weaker even disappears in some cases. However weak two-magnon feature in underdoped HTSC demonstrates a novel and unexpected behavior under an applied magnetic field. Figure 20.14 shows two-magnon Raman peaks in LSCO ($x = 0.12$) with and without an applied field [20]. One can see that at low temperatures (below $T_c$) the weak two-magnon peak is strongly enhanced by a high magnetic field of 14 T, which implies that AF correlation is strengthened largely. It is a really surprising result. Coincidentally, from neutron scattering experiments one can find that AF resonance mode at $(\pi, \pi)$ is also largely enhanced when a high magnetic field of 14.5 T is applied [21], which is well consistent with two-magnon Raman scattering. Both experiments clearly reveal that AF spin fluctuations can be strongly enhanced by applied field in superconducting state. This is considered to be related to magnetic vortices driven by external magnetic field in superconducting state. Some theoretical models state that in a vortex core normal state is recovered, as a consequence it contributes to strong AF signal in two-magnon Raman and neutron scattering experiments.

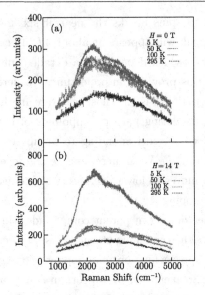

Fig. 20.14   The temperature and magnetic field dependence of two-magnon peak intensity in the underdoped LSCO. The two-magnon peak is dramatically boosted up under an applied field of 14 T in the superconducting state. Reprinted with permission from [20]. Copyright (2005) by the American Physical Society.

## 20.4   Overview

Raman scattering is the response to the interaction between light and elementary excitations in material, so it can detect and provide plenty of information on quasiparticles such as electron, optical phonon, magnon, exciton, plasmon, etc. Moreover, one can use the unique advantage of symmetry analysis to study anisotropic problems, which is particularly useful for $d$-wave superconductors. And we can obtain precise superexchange interaction by two-magnon process. Raman scattering has become one of the most fundamental experimental techniques in studying HTSC.

So far in the study of HTSC, Raman scattering has made important progress and consensuses have been reached in some aspects, such as $d$-wave pairing symmetry, superconducting gap and its doping dependence, AF correlation in two-magnon process, anisotropic properties of nodes and antinodes and so on. But there are still some problems that have not been well understood:

(1) There is a finite background in Raman spectra up to 1 eV which remains unknown till now.

(2) Two-gap observations put forward a question on the origin of $A_{1g}$ and $B_{1g}$ contributions again.

(3) Pseudogap problem in the normal state in Raman scattering.

(4) The evolution of the pair-breaking peak from overdoped to underdoped state needs to be further investigated.

(5) The evolution of two-magnon peak with doping and its relation to AF spin fluctuations.

Raman scattering plays an important role in studying HTSC, meanwhile the study of HTSC in return significantly contributes to the rapid developments of Raman scattering in both technique and theory, particularly in electronic Raman scattering. In the future, Raman scattering with extreme sample environments such as high pressure and magnetic field will provide us with more opportunities to reveal new phenomena, which will help us establish deeper insight into HTSC mechanism.

This work was supported by the 973 program under Grant No.~2011CBA 00112, by the NSF of China under Grant Nos.~11034012 and 11004243, by the Fundamental Research Funds for Central Universities, and by the Research Funds of Renmin University.

[1]   X. Tao, *d-wave supercondutor* (Sicence Press, Beijing, 2007).

[2]   T. P. Devereaux, R. Hack, Rev. Mod. Phys. **79**, 175 (2007).

[3]   F. Venturini, M. Opel, T. P. Devereaux *et al.*, Phys. Rev. Lett. **89**, 107003 (2002).

[4]   H. L. Liu *et al.*, Phys. Rev. Lett. **82**, 3524 (1999).

[5]   M. Opel *et al.*, Phys. Rev. B **61**, 9752 (2000).

[6]   J. G. Naeini *et al.*, Phys. Rev. B **59**, 9642 (1999).

[7]   F. Venturini *et al.*, Phys. Rev. B **66**, 060502(R) (2002).

[8]   M. Kang *et al.*, Phys. Rev. B **56**, R11427 (1997).

[9]   A. Sacuto *et al.*, arXiv:cond-mat/ 9804060.

[10]   B. Stadlober *et al.*, Phys. Rev. Lett. **74**, 4911 (1995).

[11]   M. M. Quazilbash *et al.*, Phys. Rev. B **72**, 214510 (2005).

[12]  G. Blumberg *et al.*, Phys. Rev. Lett. **88**, 10702 (2002).

[13]  M. Le Tacon *et al.*, Nature Physics **2**, 537 (2006).

[14]  H. Martinho *et al.*, Phys. Rev. B **69**, 180501(R) (2004).

[15]  R. Zeyher and A. Greco, Phys. Rev. Lett. **89**, 17004 (2002).

[16]  G. Blumberg *et al.*, Science **278**, 1427 (1997).

[17]  F. Venturini *et al.*, Phys. Rev. Lett. **89**,107003 (2005).

[18]  L. Tassini *et al.*, Phys. Rev. Lett. **95**, 117002 (2005).

[19]  M. Rubhausen *et al.*, Phys. Rev. B **56**, 14797 (1997).

[20]  L. H. Machtouh, B. Keimer and K. Yamada, Phys. Rev. Lett. **94**, 107009 (2005).

[21]  B. Lake *et al.*, Nature **415**, 299(2002).

# Index